# top SHELF
## BIOLOGY

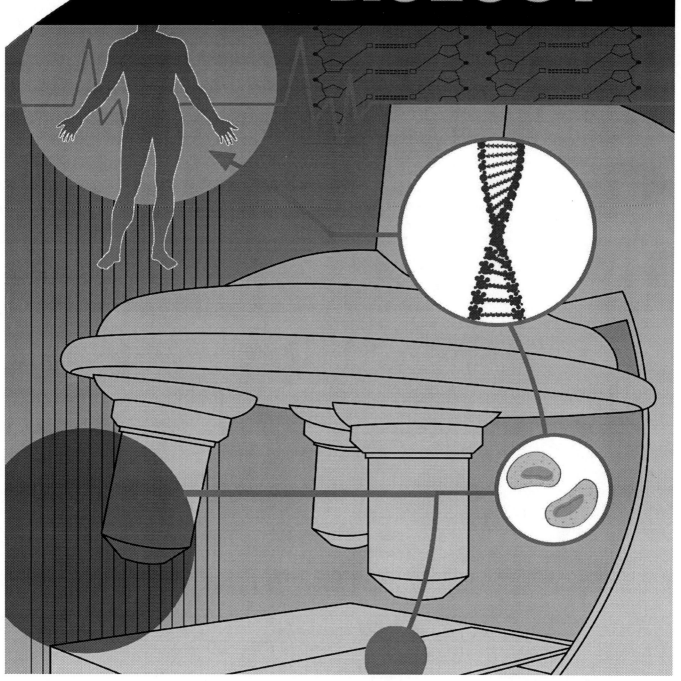

Gina Hamilton

**Acknowledgments**

We would like to thank the National Oceanic and Atmospheric Administration (NOAA), the Bureau of Land Management (BLM), the Department of the Interior (DOI), and the National Park Service (NPS) for the illustrations and photographs that appear throughout this book.

# User's Guide
# to
# Walch Reproducible Books

Purchasers of this book are granted the right to reproduce all pages.

This permission is limited to a single teacher, for classroom use only.

Any questions regarding this policy or requests to purchase further reproduction rights should be addressed to

Permissions Editor
J. Weston Walch, Publisher
321 Valley Street • P.O. Box 658
Portland, Maine 04104-0658

1 2 3 4 5 6 7 8 9 10

ISBN 0-8251-4624-0

Copyright © 2003
J. Weston Walch, Publisher
P.O. Box 658 • Portland, Maine 04104-0658
walch.com

Printed in the United States of America

# Contents

**PREFACE**
- National Science Standards for High School . . . . . . . . . . . . . . . . . . . . . . . . . . . *vi*
- Safety and Ethical Issues . . . . . . . . . . . . . . . . . . . . . . . . . . . . . . . . . . . . . . . . . . *vii*
- Parent/Teacher/Student Guide . . . . . . . . . . . . . . . . . . . . . . . . . . . . . . . . . . . . . *viii*

**BIOLOGY BACKGROUND**
- What Is Biology? . . . . . . . . . . . . . . . . . . . . . . . . . . . . . . . . . . . . . . . . . . . . . . . . . 1

**ECOLOGY**
- What Is Ecology? . . . . . . . . . . . . . . . . . . . . . . . . . . . . . . . . . . . . . . . . . . . . . . . . 5
- Symbiosis . . . . . . . . . . . . . . . . . . . . . . . . . . . . . . . . . . . . . . . . . . . . . . . . . . . . . . 8
- Student Lab: Symbiotic Lichen . . . . . . . . . . . . . . . . . . . . . . . . . . . . . . . . . . . . 11
- Food Chains . . . . . . . . . . . . . . . . . . . . . . . . . . . . . . . . . . . . . . . . . . . . . . . . . . . 12
- Larger Patterns in Ecosystems . . . . . . . . . . . . . . . . . . . . . . . . . . . . . . . . . . . . 15

**MONERA AND PROTOCTISTA**
- Kingdom Monera . . . . . . . . . . . . . . . . . . . . . . . . . . . . . . . . . . . . . . . . . . . . . . 18
- Student Lab: Bacterial Shapes and Arrangement . . . . . . . . . . . . . . . . . . . . . 21
- Moneran Structure and Reproduction . . . . . . . . . . . . . . . . . . . . . . . . . . . . . . 22
- Monerans and Disease . . . . . . . . . . . . . . . . . . . . . . . . . . . . . . . . . . . . . . . . . . 25
- Moneran Adaptation . . . . . . . . . . . . . . . . . . . . . . . . . . . . . . . . . . . . . . . . . . . 28
- Kingdom Protoctista . . . . . . . . . . . . . . . . . . . . . . . . . . . . . . . . . . . . . . . . . . . 31
- Protozoans . . . . . . . . . . . . . . . . . . . . . . . . . . . . . . . . . . . . . . . . . . . . . . . . . . . 34
- Algae . . . . . . . . . . . . . . . . . . . . . . . . . . . . . . . . . . . . . . . . . . . . . . . . . . . . . . . . 37
- Slime Molds and Water Molds . . . . . . . . . . . . . . . . . . . . . . . . . . . . . . . . . . . 40
- Student Lab: Growing Water Molds . . . . . . . . . . . . . . . . . . . . . . . . . . . . . . . 42

**EUKARYOTIC LIFE**
- Eukaryotic Cell Structure . . . . . . . . . . . . . . . . . . . . . . . . . . . . . . . . . . . . . . . . 43
- Heredity and Genetics . . . . . . . . . . . . . . . . . . . . . . . . . . . . . . . . . . . . . . . . . . 46
- Ploidy, Meiosis, and Mitosis . . . . . . . . . . . . . . . . . . . . . . . . . . . . . . . . . . . . . 50
- Student Lab: Observing Cell Division in Onions . . . . . . . . . . . . . . . . . . . . . 54

## FUNGI AND PLANTS

- Kingdom Fungi . . . . . . . . . . . . . . . . . . . . . . . . . . . . . . . . . . . . . . . . . 55
- Zygospores and Sac Fungi . . . . . . . . . . . . . . . . . . . . . . . . . . . . . . . . 59
- Club Fungi and the Deuteromycotes . . . . . . . . . . . . . . . . . . . . . . . . 61
- Fungi in the Nutrient Cycles . . . . . . . . . . . . . . . . . . . . . . . . . . . . . . 63
- Student Lab: Bread Mold . . . . . . . . . . . . . . . . . . . . . . . . . . . . . . . . 66
- Kingdom Plantae . . . . . . . . . . . . . . . . . . . . . . . . . . . . . . . . . . . . . . 67
- Photosynthesis . . . . . . . . . . . . . . . . . . . . . . . . . . . . . . . . . . . . . . . . 71
- Spore-Bearing Plants . . . . . . . . . . . . . . . . . . . . . . . . . . . . . . . . . . . 74
- Spores and Seeds . . . . . . . . . . . . . . . . . . . . . . . . . . . . . . . . . . . . . . 77
- Gymnosperms . . . . . . . . . . . . . . . . . . . . . . . . . . . . . . . . . . . . . . . . 80
- Student Lab: Tree Rings . . . . . . . . . . . . . . . . . . . . . . . . . . . . . . . . . 83
- Angiosperms . . . . . . . . . . . . . . . . . . . . . . . . . . . . . . . . . . . . . . . . . 84
- Roots, Stems, and Leaves . . . . . . . . . . . . . . . . . . . . . . . . . . . . . . . . 88
- Flowers and Fruits . . . . . . . . . . . . . . . . . . . . . . . . . . . . . . . . . . . . . 92

## ANIMALS

- Kingdom Animalia . . . . . . . . . . . . . . . . . . . . . . . . . . . . . . . . . . . . . 95
- Marine and Aquatic Invertebrates . . . . . . . . . . . . . . . . . . . . . . . . . . 99
- Terrestrial Invertebrates . . . . . . . . . . . . . . . . . . . . . . . . . . . . . . . . 104
- Arthropods . . . . . . . . . . . . . . . . . . . . . . . . . . . . . . . . . . . . . . . . . 107
- Arthropod Evolution and Diversity . . . . . . . . . . . . . . . . . . . . . . . 111
- Sea Squirts and Invertebrate Chordates . . . . . . . . . . . . . . . . . . . . 115
- The Vertebrate Classes . . . . . . . . . . . . . . . . . . . . . . . . . . . . . . . . 117
- Fish . . . . . . . . . . . . . . . . . . . . . . . . . . . . . . . . . . . . . . . . . . . . . . 119
- Student Lab: How Fish Swim . . . . . . . . . . . . . . . . . . . . . . . . . . . 123
- Amphibians . . . . . . . . . . . . . . . . . . . . . . . . . . . . . . . . . . . . . . . . 124
- The Amniotic Egg . . . . . . . . . . . . . . . . . . . . . . . . . . . . . . . . . . . 128
- Reptiles . . . . . . . . . . . . . . . . . . . . . . . . . . . . . . . . . . . . . . . . . . . 131
- Birds . . . . . . . . . . . . . . . . . . . . . . . . . . . . . . . . . . . . . . . . . . . . . 136
- Student Lab: Feathers . . . . . . . . . . . . . . . . . . . . . . . . . . . . . . . . . 139
- Mammals . . . . . . . . . . . . . . . . . . . . . . . . . . . . . . . . . . . . . . . . . 140

**EVOLUTION**

Defining Evolution . . . . . . . . . . . . . . . . . . . . . . . . . . . . . . . . . . . . . . 147

Student Lab: Creating "Life" . . . . . . . . . . . . . . . . . . . . . . . . . . . . . . 150

Precambrian Time . . . . . . . . . . . . . . . . . . . . . . . . . . . . . . . . . . . . . . 152

The Paleozoic Era . . . . . . . . . . . . . . . . . . . . . . . . . . . . . . . . . . . . . 155

The Mesozoic Era . . . . . . . . . . . . . . . . . . . . . . . . . . . . . . . . . . . . . 159

The Cenozoic Era . . . . . . . . . . . . . . . . . . . . . . . . . . . . . . . . . . . . . 163

Hominids . . . . . . . . . . . . . . . . . . . . . . . . . . . . . . . . . . . . . . . . . . . 166

Human Evolution: *Homo* . . . . . . . . . . . . . . . . . . . . . . . . . . . . . . . . 168

**APPENDIX I**

Answer Key . . . . . . . . . . . . . . . . . . . . . . . . . . . . . . . . . . . . . . . . . 171

Rubrics: Assessing Laboratory Reports . . . . . . . . . . . . . . . . . . . . . . . 179

Rubrics: Assessing Essays . . . . . . . . . . . . . . . . . . . . . . . . . . . . . . . . 180

**APPENDIX II**

Scientific Supplies and Suppliers . . . . . . . . . . . . . . . . . . . . . . . . . . . 181

**APPENDIX III**

A Time Line of Biology . . . . . . . . . . . . . . . . . . . . . . . . . . . . . . . . . 183

**GLOSSARY** . . . . . . . . . . . . . . . . . . . . . . . . . . . . . . . . . . . . . . . . . . . . 185

**INDEX** . . . . . . . . . . . . . . . . . . . . . . . . . . . . . . . . . . . . . . . . . . . . . . 187

# National Science Standards for High School

The goals for school science that underlie the National Science Education Standards are to educate students who are able to

- experience the richness and excitement of knowing about and understanding the natural world;

- use appropriate scientific processes and principles in making personal decisions;

- engage intelligently in public discourse and debate about matters of scientific and technological concern; and

- increase their economic productivity in their careers through using knowledge, understanding, and skills they have acquired as scientifically literate individuals.

These goals define a scientifically literate society. The standards for content define what the scientifically literate person should know, understand, and be able to do after 13 years of school science. Laboratory science is an important part of high-school science, and to that end, we have included several labs in each volume of *Top Shelf Science*.

The four years of high-school science are typically devoted to earth and space science in ninth grade, biology in tenth grade, chemistry in eleventh grade, and physics in twelfth grade. Students between grades 9 and 12 are expected to learn about modeling, evidence, organization, and measurement, and to achieve an understanding of the history of science. They should also accumulate information about scientific inquiry, especially through laboratory activity.

Our series, *Top Shelf Science,* addresses not only the national standards, but also the underlying concepts that must be understood before the national standards issues can be fully explored. National standards are addressed in specific tests for college-bound students, such as the SAT II, the ACT, and the CLEP. We hope that you will find the readings and activities useful as general information as well as in preparation for higher-level coursework and testing. For additional books in the *Top Shelf Science* series, visit our web site at walch.com.

# Safety and Ethical Issues

The *Top Shelf Science* series contains several laboratory experiments. Special care must be taken to ensure student safety when performing these experiments. Experiments involving living organisms should be done with careful respect for the health of the living specimen in mind. Here are some guidelines for general safety issues in a laboratory setting:

- Wear proper safety equipment at all times. This includes an apron, smock, or lab coat; safety goggles; and gloves. Do not wear open-toed shoes, such as sandals, during lab experiments.

- Do not eat or drink anything in the lab.

- Be sure to turn off heat sources when not in use.

- Perform any chemical experiments involving gas emissions within a chemical fume hood or in a well-ventilated room.

- Before disposing of chemical ingredients, be certain that they are neutralized, then dispose of them in proper containers.

- Establish a location for the disposal of sharp objects, such as broken glass or nails.

- Use extreme caution when heating solutions.

- Animals, plants, and other life forms are deserving of respect. Treat living specimens with care and, when possible, release them or replant them outdoors.

- Use care when using electrical appliances of any sort. Know how to recognize a short circuit or a blown fuse.

- Keep fire extinguishers on hand and properly charged, and know how to use them. Be sure that you have an ABC-rated extinguisher, as well as a Halon™ extinguisher for electrical fires.

- Follow all local, state, and federal safety procedures.

- Have evacuation plans clearly posted, planned, and actually tested.

- Label all containers and use original containers. Dispose of chemicals that are outdated.

- Be especially aware of the need to dispose of hazardous materials safely. Some chemistry experiments create by-products that are harmful to the environment.

- Take appropriate precautions when working with electricity. Make sure hands are dry and clean, and never touch live wires, even if connected only to a battery. Never test a battery by mouth.

- When using lasers, never look directly into the beam, and make sure students are conversant with the dangers of laser light.

Safety precautions unique to a given laboratory will be provided within the lab write-up itself. These safety precautions are provided as a guide only. They may be incomplete. Use common sense when working with any chemicals, electricity, or living organisms.

# Parent/Teacher/Student Guide

Dear Parents, Teachers, and Students,

Thank you for choosing the *Top Shelf Science* series to help you better understand some of the difficult ideas in high-school science. We are confident that our books will help students who have a greater knowledge of the subject matter being studied; they can also be used to provide a lab-based program for students learning at home.

Each volume of the *Top Shelf Science* series is designed for a particular course of study. Within each volume, concepts build sequentially, and it is recommended that students begin with the first section and move forward.

Each book has sections that are thematically designed. The laboratory exercises associated with each section are specific to a deeper understanding of the overlying concept. In Appendix II, you will find a list of materials that are necessary to conduct the lab exercises, as well as a list of science equipment dealers who may help you acquire things you need in the course of the lab exercise; we have tried to keep the materials list small, as well as provide lab lessons in which materials are readily accessible. Therefore, we have also provided alternatives, where possible, to the lab glassware and other large pieces of equipment that may not be located in your kitchen cabinet or small classroom.

In Appendix I, you will find answers to the questions in each unit, as well as a suggested grading rubric for essays and lab reports. Share these rubrics with students so that they can correct areas that need to be corrected before the next assignment. In keeping with the national science standards, we have also included a time line of the history of each discipline. Each volume also contains an index and a glossary.

Whether you are using our product as the basis for a home school experience, a new and fresh way of supporting textbook material, or as preparation for a college placement test, we are confident that *Top Shelf Science* can meet your needs.

Thank you!

The authors and editors of *Top Shelf Science*

## What Is Biology?

Biology is the study of living organisms, their functions, their interactions, their evolution, and their *taxonomy,* or classification.

The currently accepted model of taxonomy of living things includes five large kingdoms of life—monera, protoctista, fungi, plants, and animals. Each of these kingdoms is broken down further into phylum, class, order, family, genus, and species. (An easy way to recall the order of classification is to remember this simple mnemonic: King Phillip Came Out From Germany Singing.) In general, when a species is named, only the genus and species are listed. It is convention that the genus is capitalized, while the species name is not—for example, *Homo sapiens.* However, each species also belongs to all the other classifications. It is a little like having a name with seven separate words, and using only the last two "socially." Our own species, *Homo sapiens,* is also a member of the kingdom Animalia, the phylum Chordata, the class Mammalia, the order Primate, the family Hominidae, and the genus and species *Homo sapiens.*

The classification system in use today was devised by Carolus Linnaeus, a Swedish botanist, in the eighteenth century. His system, unlike earlier ones, was based on shared characteristics of organisms. The more characteristics they shared, the more closely they were related. For instance, all mammals possess hair, nourish their offspring on milk produced by the mother, and are endothermic with a four-chambered heart. It doesn't matter whether you are an opossum or a caribou, these characteristics apply to you. As characteristics diverge—for instance, opossums and caribou are in two different orders—they are more closely related to species within the order than they are to each other. Not only are animals classified this way, but all living things in the five kingdoms of life are so classified. The Linnaean system gives us a very neat, orderly view of how closely or distantly life forms are related.

Over the years, there have been many other ideas about classification. Until recently, only two kingdoms were recognized—plants and animals. Under this scheme, all organisms capable of photosynthesis, and some that were decomposers, were considered

> **Biology is the study of living organisms, their functions, their interactions, their evolution, and their *taxonomy,* or classification.**

"plants," while all organisms that consumed them were considered "animals." Some of the names of organisms still reflect this older idea. Some biologists still believe that, in general, organisms should be defined in terms of their part in the ecosystem—producers (such as photosynthetic organisms), consumers, and decomposers. Others believe that the five-kingdom model should be expanded into six kingdoms to include two kingdoms of bacteria.

Even the Linnaean classification system is under review. While the Linnaean system addresses shared characteristics, as more of the fossil record becomes known, it is possible to actually address shared common ancestors. A new system for classifying organisms—cladistics—shows how organisms adapted and changed over time.

> **Viruses are in the twilight between life and non-life.**

You will notice that viruses are not included in the five-kingdom model. Why? The definition of a living organism is one that has *DNA* (deoxyribonucleic acid, strands of genetic material that carry the code for one's hereditary properties, such as eye color), and the ability to metabolize, grow, reproduce at the adult level, and react to stimuli. Viruses are missing several key pieces to the puzzle. Viruses do not grow, and they can only reproduce when they are within another organism, using that organism's metabolic capability to do it—they do not possess the enzymes that true living things possess to carry out life functions. They are always parasitic in nature. Viruses can be extremely damaging to living organisms, often killing their host.

What is a virus, then, if it is not technically alive? Viruses are tiny (about 100 nm long), extremely complex molecules that have some sort of nucleic acid—usually *RNA* (ribonucleic acid), but occasionally DNA—that carries its genetic code, a coat of protein that protects the molecule, and a lipid (or fatty) membrane that surrounds the protein coat. They are like tiny machines, possessing one, perhaps two, enzymes that allow them to reproduce in the cell of a living thing. They are in the twilight between life and non-life.

Viruses reproduce by invading host cells, "reprogramming" the host cell's DNA with the virus's own RNA or DNA, replicating and creating new virus particles within the cell, and finally destroying the cell in order to release the new virus particles. Each new *virion*, or virus particle, is then free to invade other cells.

> Viruses cause human diseases such as the common cold, chicken pox, smallpox, mumps, measles, AIDS, and polio.

Viruses, unlike bacterial infections, cannot be treated successfully with antibiotics. There are a few anti-viral drugs that work on a limited basis, but most viral diseases are prevented, with vaccination, rather than being treated after infection. Viruses cause human diseases such as the common cold, chicken pox, smallpox, mumps, measles, AIDS, and polio.

How did viruses arise? Although they seem very simple, viruses cannot have occurred before true living things arose in the evolutionary timescale. Since viruses require host cells for their reproduction (and most are species-specific), they are probably mutant organisms that originated from the host cells themselves. It is believed that at some point in viruses' evolution, nucleic acids escaped from host cells and developed a way to reproduce by parasitizing other host cells of the same species. If this hypothesis turns out to be true, host cells and the viruses that parasitize them are more closely related to each other than different viruses are to one another. Sadly, this also means that new viruses can and do arise regularly, as we have learned with seemingly new viruses, such as AIDS and the Hanta virus.

 **Exploration Activities**

1. What is the order of classification?

2. Why is a virus not technically a living organism?

3. Describe the reproduction of a virus.

## What Is Ecology?

An understanding of how animals, plants, and other organisms relate to one another and the nonliving parts of their natural world is the domain of ecology. The word *ecology* means "the science of the home" because this part of life science is about the organism in its environment.

There are several levels of organization in ecology. The largest ecological community in the biosphere (all areas where life can exist on Earth), is a **biome.** Biomes are major divisions in the ecological world—for instance, grasslands, rainforests, tundras, deserts, seashores, or deep oceans. Within the biome, there may be several smaller **ecosystems** with different populations of organisms, even though they are in the same biome. An isolated tidepool, for instance, is one ecosystem within the larger seashore biome. An ecosystem is any complete environment that contains both *abiotic* (nonliving) and *biotic* (living) factors, and functions more or less independently.

An abiotic factor is anything that affects the ecosystem but is not alive. For instance, rainfall patterns, soil nutrients, high and low tides, and the amount of sunlight are all abiotic factors. Sometimes abiotic factors limit the number of individuals that a region can support. Temperature, for instance, in the Arctic, is such a **limiting factor.** Since no living thing can survive and grow without water, rainfall is often a limiting factor in ecosystems.

Biotic factors are anything that is or was alive—bacteria, protoctists, fungi, plants, and animals. Often, these factors will be split into two categories—*flora,* which means anything plant-like that is capable of becoming a food producer in the ecosystem, and *fauna,* which is the animal kingdom and any other consumers. Historically, photosynthesizing bacteria and algae, and most fungi, were grouped into the flora category.

The interactions of flora and fauna in an ecosystem are studied as part of ecology to understand the driving force in that particular ecosystem. There is always competition—for food, shelter, and mating. There is competition within species and between species. This competition governs how the species and the ecosystem itself

> An ecosystem is any complete environment that contains both *abiotic* (nonliving) and *biotic* (living) factors, and functions more or less independently.

Tidepool, home to thousands of organisms

evolve. But within each ecosystem, there are also built-in mechanisms for sharing the available resources among species. Some species of trees grow taller than others, allowing all the trees to receive sunlight. Some small animals eat seeds, while others eat only fruit. Each species thrives on specific factors—amount of sunlight needed, specialized food requirements—within the larger ecosystem. These specific factors form an ecological niche and allow many species to live within the same ecosystem more or less harmoniously.

Humans, more than any other species, are capable of changing entire biomes. Agriculture, maritime industries, logging, and many other human activities have the capability of degrading the environment. Scientists who observe specific ecosystems and biomes often determine the health of the communities based on *indicator species,* animals, usually fairly high in the food chain, whose presence or absence tells us about the numbers and health of the rest of the species in the community. Long ago, miners used to take a canary in a cage down into the coal mines. If the canary became ill or died, the miners knew the air and environment in the mine was too dangerous for them, and they got out. In the same way, the indicator species tell ecologists about the effects of logging, pollution, and other human activities on the habitat in which the indicator species lives.

> **Humans, more than any other species, are capable of changing entire biomes.**

It is true, and sometimes sad, that the ecosystems of greatest diversity—the largest number of species per unit of land or water—have the smallest number of individuals of any one species within it. Kill a couple of corn plants in a cornfield, and the cornfield is virtually unchanged. But if a couple of rare orchids in a rainforest die, they may be among the last of their species, and the world is forever diminished by their loss.

Ecology, then, is the big picture in biology. By looking at how all organisms and the abiotic factors in particular communities interact, we learn more about each individual species.

 **Exploration Activities**

1. Rank the following from the largest number of organisms served to the smallest number of organisms served: biome, niche, ecosystem, biosphere.

2. Name one ecosystem within a grassland biome.

3. What is the difference between abiotic and biotic factors?

4. Define the terms flora and fauna.

5. What is an indicator species?

6. What is diversity?

# Symbiosis

Within particular ecosystems, competition for available resources results in interactions between species. Together, all interactions and relationships among species are known as *symbiosis,* or "living together." Many relationships, even those that are harmful to individuals, have an overall benefit to the ecosystem at large. However, when we consider the various symbiotic relationships, they are always individual relationships—between individual lion and individual gazelle, between individual bacterium and individual host, between individual vine and individual tree.

Some relationships are beneficial for both parties, others are neutral for one party and beneficial for the other, while some harm one party but are beneficial for the other. Each of these different forms of symbiosis has a different name.

One such relationship is **parasitism.** As the name implies, parasitic relationships occur when one party in the relationship is harmed, but the other is benefited. Most predator/prey relationships are parasitic, even if they are healthy for the ecosystem (and even the population of the prey species) as a whole. Another kind of parasitic relationship is a more long-term relationship, in which one organism, over time, uses another organism and does harm to it. When a tick attaches itself to an animal's skin, the tick is enjoying a constant source of food—the animal's blood. However, the depletion of blood is harmful to the host animal. In addition, ticks sometimes carry and transmit diseases that can harm, or even kill, the host. Parasitic relationships occur between animals, as our example shows, but they also occur within and among every kingdom of life. Clinging vines that ultimately kill the tree they are using to reach sunlight, bacterial infections that harm their animal or plant host, and fungi that spread via particular plant species are all parasitic relationships in nature.

Another symbiotic relationship is **commensalism.** Commensal relationships occur when only one party benefits from a relationship, and neither party is harmed. The word *commensal* means "at the table together." Scavengers such as buzzards, who clean up after a lion eats its dinner, benefit from their relationship with the lion.

> Together, all interactions and relationships among species are known as *symbiosis,* or "living together."

The lion, who gets no direct benefit, is not harmed in the exchange. Many plants and other species also have commensal relationships, for example, harmless bacteria that live within our own digestive systems, and algae that attach to the backs of mollusks.

> In mutualism, both parties benefit.

The type of symbiotic relationship most people think of when they think about symbiosis is the mutual relationship. In **mutualism**, both parties benefit. The Egyptian plover, a small bird, can often be seen cleaning the teeth of the Nile crocodile. The crocodile does not harm the bird, and the bird gets scraps from the crocodile's teeth to eat. Just as with the other types of symbiosis, all kingdoms of life participate in mutual relations. The largest obvious relationship is the relationship of the entire plant kingdom (together with some bacteria and protoctists) with the entire animal kingdom. Photosynthesizing plants, bacteria, and algae provide oxygen that is used by the animal kingdom, along with other organisms. The waste product of metabolizing oxygen is carbon dioxide, which is, in turn, used by the plants for their own photosynthesis. Without this dynamic relationship, life would not be possible on Earth.

 **Exploration Activities**

Decide whether the relationships below are parasitic, mutual, or commensal.

1. A small African antelope, the dik-dik, uses abandoned anteater nests, which are located underground, to hide from large predators, such as lions and hyenas.

2. A tapeworm attaches itself to the stomach lining of a wolf, absorbing many of the nutrients the wolf consumes.

3. The flashlight fish, which lives in deep water off the coast of Lebanon, has pockets under its eyes. A bioluminescent bacteria lives in the pockets and emits a soft intermittent light, allowing the fish to see in the dark water.

# Student Lab: Symbiotic Lichen

Lichen can often be seen growing on trees. They appear to be grayish green. They exhibit a symbiotic relationship. Can you tell what kind of symbiosis they exhibit from this investigation?

## Materials

- Fruticose lichen, which can be found on trees
- Microscope, slide, and cover slip

## Special Safety Consideration

While lichens are generally harmless, many fungi are poisonous. Be sure to wash your hands thoroughly after handling any fungus.

## Procedure

Crush a small piece of the lichen (about 2 centimeters) between your fingers and make a wet mount of lichen on a slide. Focus on the drop under low power. Increase the magnification until you see a brighter green area. When you have located a region, switch to high power. What do you see?

## Conclusion

When you look at the slide, you should see some green spheres and gray fragments. The green spheres are *cyanobacteria,* photosynthesizing members of the Kingdom Monera. The strings are fungi. A lichen only exists because the bacteria and the fungi live in a mutual relationship. The bacteria provide food for the fungi, and the fungi allow the bacteria to live within their **hyphae,** providing a protected environment. When separated in the laboratory, the fungi always die, but the bacteria can survive on their own.

## Food Chains

In any ecosystem, there are those who make the food, those who eat it, and those who clean up after dinner. One way to look at food chains in the natural world is to think of them as energy transfers.

The organisms that provide the energy are called the *producers*. These are mostly plants and algae, along with some photosynthetic microorganisms. They create simple sugars through the process of photosynthesis. Simply put, photosynthesizing organisms use carbon dioxide and water, in the presence of sunlight, to create simple sugars that plants and other photosynthesizing organisms use for their own growth and reproduction. In the section on plants, we will show how photosynthesis works in detail. Plants, in their cell walls, also contain certain vitamins that all living things require. They also take up minerals, such as iron, from the soil.

Animals that eat the plants and other organisms are called *first-order consumers*. The energy built up by the plants is transferred down the line to *herbivores* (animals that eat only plants). The simple sugars are built into more complex nutrients such as fats, proteins, and more complex carbohydrates within the bodies of the first-order consumers. At each transference, a little energy is lost in the exchange through heat loss of the animals involved. The sun continues to replenish the producers.

In turn, first-order consumers pass their energy along to *second-order consumers,* who eat them. An example of a food chain would be a gazelle that consumes grass and is in turn eaten by a lion. In this chain, the grass is the producer species, the gazelle is the first-order consumer, and the lion is the second-order consumer. Second-order consumers are usually *carnivores* (animals that eat only meat), although some eat both plant and animal matter. Animals that consume both plants and animals are called *omnivores,* and they are, variously, first-order consumers and second-order consumers. Some animals are *insectivores*. As the name implies, they eat insects, who survive mainly on plant nutrition. Insectivores and animals that survive on eating the eggs of other species are also second-order consumers.

*Third-order consumers* are predators who eat other predators. For instance, a hawk that eats a snake, which has just eaten a mouse, which survives on seeds, is a third-order consumer.

The next link in the food chain belongs to the scavengers who feed directly on dead or dying animal tissue. The buzzard that finishes the lion's gazelle kill and the coyote that eats a squirrel that died of natural causes are both scavengers.

The last link in any food chain is decomposition. Many organisms, such as bacteria, small invertebrates, and fungi, reduce dead plant or animal tissues to the molecular level, where they can be processed in the soil and used to build new plant and animal tissue. At this final level, energy is totally lost, but nutrients are returned to the soil, where they can be used by plants to build their structures to begin the cycle anew.

> **The last link in any food chain is decomposition.**

Scavengers and decomposers play a vital role in all ecosystems. Without them, dead animal and plant tissue would remain in the environment. Imagine what your yard would look like after just a couple of years of leaf-fall if none of the leaves decomposed!

How far the energy has traveled from the producer level through the consumer orders is referred to as the *trophic level*. The consumption of seeds by a mouse is the first trophic level, just as the mouse is a first-order consumer. As the mouse is consumed by a snake, the energy is another step removed from its producer. Plants and other producers are called *autotrophs,* while organisms that feed on them at any trophic level are *heterotrophs*.

Animals typically have a varied diet, with even first-order consumers eating many species of plants. Food chains, therefore, become more of an interconnected web of energy transfers. Food webs are a more realistic view of how energy is transferred from producers through consumers and scavengers, and on to decomposers.

 **Exploration Activities**

1. Design a food chain that includes the following organisms: fruit finch, apple tree, bobcat, coyote, and bacteria.

2. What is the importance of decomposers in the food chain?

3. Describe the difference between a food chain and a food web.

4. What is the difference between an autotroph and a heterotroph?

**BACKGROUND**

# Larger Patterns in Ecosystems

In addition to abiotic factors, flora, and fauna, there are often large-scale patterns that affect how the ecosystem thrives. Each member of the system is affected, positively or negatively, based on these large changes to the system. Over time, we can see how interconnected each ecosystem is and, sometimes, how fragile or resilient an ecosystem is.

What are these large-scale patterns? Seasonal variability, weather or climate change, migration patterns, tidal variability, and many others.

For instance, in the African grasslands, life is dominated by two great seasons—the brief wet season, when the grass becomes lush, leaves on the trees swell, the watering holes and rivers fill, and seasonal springs come to life; and the dry season, when grass grows golden brown, leaves become tough, and water dries up. Animals live their lives by these seasons, giving birth to their young in the rainy times and migrating to cooler and wetter upland areas as the dry season drags on.

A very small, temporary change to such an ecosystem can cause widespread upheaval throughout its populations. If the rains come a few days late, or fail to come at all during the short rainy season, most of a generation of plants will fail to germinate, and a whole generation of animals will either not be born at all, or will die shortly after birth. This has repercussions throughout the food chain for several years to come. Usually, the balance is eventually restored.

The most dramatic changes, however, occur over long periods of time. Long-term drought, for instance, not only affects biotic factors in the ecosystem, but abiotic factors as well. The Sahara, for example, is creeping southward at a slow but steady rate, which changes weather patterns in the region and causes even greater desertification. The changes to the abiotic system—average temperature increase and rainfall decrease—have a dramatic impact on the biotic system in loss of producer species, first-order consumers, and the second-order consumers. A previously thriving ecosystem shrinks.

> **A very small, temporary change to an ecosystem can cause widespread upheaval throughout its populations.**

> **Ice ages are fairly regular occurrences in Earth's history.**

An example of long-term climatic change that affects all ecosystems is an ice age. Ice ages are fairly regular occurrences in Earth's history, and they have fundamental effects on the populations of plants, animals, and other organisms. During the last ice age, sea levels dropped and exposed land bridges in various places around the affected parts of the world. This changed migration for many species, including humans, who migrated from Asia to the North American continent via the Bering Land Bridge, between Alaska and Siberia. Interbreeding occurred between some species, creating hybrid species that formerly did not exist. Mammal species adapted to the cold conditions by growing thick coats of fur. Many species of plants and animals died out.

Ice ages, and the resulting mass extinctions that are associated with them, are only one of many long-term climate changes that affect biotic and abiotic factors in ecosystems. Understanding these patterns and their effect on populations gives us a deeper understanding into how natural conditions drive biology and evolution.

 **Exploration Activities**

1. Name three large-scale patterns that fundamentally affect ecosystems.

2. Explain how a small, temporary change to an ecosystem can result in upheaval that can last several years.

3. How does a long-term change, such as an ice age, affect both abiotic and biotic factors in an ecosystem?

**BACKGROUND**

> The first known fossils of monera date from about 1.5 billion years after the formation of Earth.

# Kingdom Monera

The simplest life forms on Earth are monerans. Also known as *prokaryotes* (or first cells), monerans contain an outer cell membrane capable of *osmosis* (in which nutrients and wastes are passed through the membrane) and *cytoplasm* (cellular fluid) throughout which is carried DNA, the cell's genetic code. Monerans divide by *constriction*—they pinch off somewhere in the middle of the cell and form two virtually identical daughter cells. Since DNA is spread throughout the cell, there is no need for a complex reproductive procedure—the genetic code gets passed on directly in the cytoplasm.

Except for "copy errors" that can be caused by outside agents, such as radiation or mineral poisoning, each moneran is identical to its parent. For the first several billion years of evolution, only monerans lived on Earth, and they changed very slowly. The first known fossils of Monera date from about 1.5 billion years after the formation of Earth. The first moneran's DNA still exists in billions of its descendents today.

The first monerans were *anaerobic* (unable to tolerate the presence of oxygen). In the primitive Earth's atmosphere, this was not much of a problem because no free oxygen existed. Any oxygen on the planet was bound up in molecules, such as water. Many of these anaerobic organisms survived by metabolizing hydrogen and carbon dioxide to get energy. However, one of the copy errors that drove moneran evolution caused photosynthetic organisms to arise. These organisms, the *cyanobacteria,* made their own food using sunlight, carbon dioxide, and water. They gave off a deadly gas as a waste product—oxygen.

The cyanobacteria pollution problem caused the first great mass extinction on Earth. The formerly dominant anaerobes died out or became restricted to virtually oxygen-free locations where they continue to thrive today. Earth was then dominated by the *aerobes*—those that could live and thrive in the new, oxygen-rich environments in the air and in the sea.

Today, there are two major subkingdoms of Monera—the anaerobic *archaebacteria,* descended from the first cells on Earth, and all other bacteria, the *eubacteria,* or "true bacteria." Eubacteria are far more common and more complex than the archaebacteria.

Archaebacteria have only three phyla extant on Earth—the methanogens, the halophiles, and the thermoacidophiles. Methanogens produce methane from hydrogen and carbon dioxide and live in rather nasty places—the intestinal tracts of animals, swamps with low oxygen content in the water, and sewage treatment plants. They give off a foul odor that is characteristic of such places. Halophiles (salt loving) live in extremely saline locations, such as the Dead Sea and the Great Salt Lake. Salt is usually harmful to bacteria, especially in such massive quantities, but the halophiles thrive in such regions. The third phylum, the thermoacidophiles, thrive in regions of high temperature and high acid content. Thermoacidophiles would be content living in sulphur hot springs, for instance, or in the caldera of volcanoes, where the pH can be extremely low.

The eubacteria have four major phyla. Cyanobacteria, which we have already discussed, are capable of photosynthesis. Cyanobacteria in the oceans, lakes, and ponds of the world are a major source of breathable oxygen for the animal kingdom. Cyanobacteria also produce, over long periods of time, large fossil structures called *stromatolites. Gram-positive* bacteria (so named because the first members of the phylum who were identified could be stained with Gram's stain) includes many organisms that are useful to humans and some that are dangerous. The members of this phylum cause strep throat, but they also culture yogurt, ferment wine, and produce antibiotics. Proteobacteria are essential to a macro ecosystem, because they fix nitrogen in soil, allowing plants to grow. Other members of this large phylum live symbiotically in the intestinal tracts of many mammals. Still others are involved in oxidation of minerals. Finally, the spirochetes, spiral-shaped bacteria, need organic compounds to survive—they are heterotrophic. Many are parasitic, including the species that causes the sexually transmitted disease syphilis.

> **Cyanobacteria in the oceans, lakes, and ponds of the world are a major source of breathable oxygen for the animal kingdom.**

 **Exploration Activity**

Match the vocabulary word to the correct definition.

| | |
|---|---|
| Methanogens | Found in places with high salt content |
| Proteobacteria | The oldest living organisms on Earth |
| Spirochetes | Photosynthetic bacteria |
| Halophiles | Produce methane gas |
| Archaebacteria | True bacteria; includes all aerobic forms |
| Cyanobacteria | Spiral-shaped and heterotrophic |
| Eubacteria | Contains members who fix nitrogen in soil |

# Student Lab: Bacterial Shapes and Arrangement

Eubacteria are divided into classifications based on their shapes and on how they group themselves together. The cells themselves are shaped as spirals, spheres, or rods. They self-organize into long chains, grape-like clusters, or pairs.

For instance, the bacteria that cause the disease strep throat are called *Streptococcus*. The prefix "strepto-" means chain, and the root "coccus" means sphere. When you look at a streptococcus culture under the microscope, you see chains of small spheres.

The table below will assist you in identifying bacteria during this lab.

| Arrangement | Arrangement Prefix | Shape | Shape Name (singular) |
|---|---|---|---|
| pairs | diplo- | rods | bacilli (bacillus) |
| chains | strepto- | spheres | cocci (coccus) |
| clusters | staphylo- | spirals | spirilli (spirillum) |

## Materials

- Microscope

- Prepared slides (If you do not have a set of prepared bacteria slides, obtain a drop of pond water, stain with iodine, and cover with a cover slip.)

## Procedure

1. Examine prepared slides (or your pond-water slide). Move the slide around until you can see three different bacterial shapes.

2. Draw a representative of each type, and identify its shape and arrangement.

## Moneran Structure and Reproduction

As mentioned in the last section, monerans are extremely simple organisms. Even in their simplicity, there are major differences among the monerans in terms of structure. All monerans have a cell membrane that regulates what enters the cytoplasm, a diffuse liquid that contains the DNA of the cell. Most monerans also possess a cell wall, which surrounds the plasma membrane and supports the cell in its characteristic shape.

Some, like cyanobacteria, also have a sticky capsule that surrounds the cell wall. The sticky capsule assists the cells in clinging together in their characteristic arrangements. In cyanobacteria, this sticky gelatinous substance attracts dust, dirt, and sand. After the cells die, they sometimes fall to the floor of the ocean or pond and are covered with debris. If this occurs again and again, over time, a fossil structure called a stromatolite forms. The capsule is also difficult for white blood cells to absorb, so bacteria that possess a capsule are more likely to cause disease than those without it.

Another structure many bacteria have are *pili*—strands that extend from the cell membrane, used by cells to connect to one another.

Some bacteria have **flagella**—long, tail-like structures that assist the cell in movement.

What is more telling is what monerans *do not* have. They do not have a nucleus, or any of the membrane-bound organelles of a eukaryotic cell, such as chloroplasts. The chlorophyll manufactured by the cyanobacteria is spread throughout the cytoplasm, rather than contained in small packets. Their DNA is a single, circular **chromosome**, rather than the paired chromosomes of higher organisms.

> All monerans have a cell membrane that regulates what enters the cytoplasm, a diffuse liquid that contains the DNA of the cell.

# Moneran Structure and Reproduction

*E. coli* reproduction. *E. coli* lives symbiotically in the intestinal tracts of mammals, but can cause systemic poisoning if the bacterium gets into the bloodstream of the animal.

Most metabolic functions—growth, respiration, excretion—occur within the cytoplasm of a moneran. When it is time to reproduce, a moneran simply contracts in the middle of the cell, until the cell membrane is touching on both sides. At that point, two new cells have formed from the one cell that comprised the parent cell. The process is called *binary fission*. Binary fission occurs rapidly and can happen very often. Under ideal conditions, a bacterium can reproduce once every 20 minutes, which yields huge numbers in a very short period of time.

As soon as the cells separate, they are fully functioning "adults." Unlike more complex cells, monerans need no resting period or juvenile stage. All they must do is metabolize, grow, and reproduce again.

## Top Shelf Science: Biology  MONERA AND PROTOCTISTA  Moneran Structure and Reproduction

 **Exploration Activity**

Collect a sample of cyanobacteria from a fish tank and observe it under the microscope. Once you find the organisms on your slide, switch to high power and watch one of them reproduce.

Create a series of drawings that show the process of binary fission.

## Monerans and Disease

Monerans often live harmlessly within other organisms, sometimes, as in the case of *E. coli,* in a mutual relationship with the host. However, if too many of a particular bacterium enter the system of the host (for instance, if the host unknowingly eats meat tainted with billions of bacteria), or if the bacteria are excreting toxins, or if they enter the bloodstream instead of the intestinal tract, the host can develop an infection.

Infection does not always mean the host will become ill. Each organism has a strong defense network. In mammals, this consists of an *immune system*—a virtual army of white blood cells that engulf and consume invading organisms, eliminating the danger. Occasionally, when the host is weakened from other infections, or is very young or old, moneran invasion can overwhelm the immune system of the host and cause disease. Once established, the monerans reproduce and cause systemic infection. As anyone who has suffered through a few days of food poisoning is aware, the entire body reacts to the infection, with fever, vomiting, diarrhea, great thirst, and other methods used by the body to rid itself of the troublesome bacteria or toxins.

When bacteria release toxins, as they do in the case of food poisoning, the body must and does purge itself as quickly as possible. However, other bacterial infections, while they cause pain and fever and more serious problems to the internal organs and glands, do not cause the systemic response that toxic bacteria cause. These quiet infections actually caused many more deaths in the years prior to the discovery of antibiotics than food poisoning did.

Diseases that have a bacterial origin include some of the world's great killers—black plague, tuberculosis, strep throat, and pneumonia. Unlike viral illnesses, most bacterial illnesses do not have a vaccine.

> **Diseases that have a bacterial origin include some of the world's great killers—black plague, tuberculosis, strep throat, and pneumonia.**

Sir Alexander Fleming, discoverer of *Penicillin notatum*

Most, however, do have a cure. In 1928, Sir Alexander Fleming discovered penicillin. Penicillin is produced by Penicillium, a fungus, typically seen as bread mold. It kills bacteria by interfering with the enzyme that links the sugar chains in the cell wall of the bacteria. This creates holes in the cell wall, which allow water into the cell. This ruptures the cell, killing it.

Fleming noticed that an airborne mold particle fell onto a dish of staphylococci bacteria. He was not interested in the mold, until he noticed that the bacteria near the mold particle began to die. His discovery would save millions of lives, but it took another 12 years to come to market. The discovery led to research in antibacterial drugs, called antibiotics, which can, today, cure all but the strongest bacterial infections.

Antibiotics have, in turn, led to the evolution of drug-resistant strains of bacteria, largely because of overuse and prescription for infections that are not bacterial in origin. It is very important that humans use antibiotics only for known bacterial infections.

 **Exploration Activities**

1. How do bacteria cause disease?

2. Who discovered penicillin? How?

3. What are the dangers of overuse of antibiotics?

## Moneran Adaptation

Monera exist almost everywhere in the world, including in environments no other living thing could possibly tolerate. They live under the Antarctic Ocean ice, along deep-sea vents where the water temperature is close to 400° C, within hot sulfur springs, and in the Dead Sea. Monera live in water, under the ground, within and on other living organisms, in the hottest deserts in the world, and in the air.

How can such simple organisms survive in such varied environments? Key to understanding moneran adaptation is understanding their methods of metabolism. Various monerans obtain nutrition from light, from simple inorganic molecules, and from complex organic molecules. Most require oxygen for respiration. These are called the *obligate aerobes*. Others are killed by oxygen—the *obligate anaerobes*. Still others can live with or without oxygen. These have two types of metabolism and can obtain their energy from respiration (aerobically) or from fermentation (anaerobically). Some have even more complex metabolisms—green sulfur bacteria, for instance, use hydrogen sulfide rather than water during photosynthesis, and produce sulfur, rather than oxygen, as a waste product.

The cell structure of the moneran can also help to protect it. Some monerans have sticky outer capsules, which often prevent them from being absorbed by immune systems. Some have pili, which can be used to cling to other individuals of the species, presenting would-be predators with a large chain or cluster of organisms, rather than one small bacterium.

Even with all of these adaptations, the environment is sometimes too hostile for survival. When this happens, many monerans can produce a structure called an *endospore*. Endospores have a hard outer covering that is resistant to drying out, boiling, and chemicals. While in the endospore form, the bacterium's metabolism slows—the bacterium enters a state of suspended animation, and it does not reproduce. When conditions become better, the endospore germinates, and the bacterium begins metabolizing, growing, and reproducing once again.

> Various monerans obtain nutrition from light, from simple inorganic molecules, and from complex organic molecules.

> One group of endospore-forming bacteria are those that produce a toxin that causes botulism, a food poisoning.

Because endospores can resist boiling, canned and bottled food must be placed in sterile containers, or endospores will survive. One way to do this is to boil the containers in water under pressure.

Have you ever watched someone make jam or pickles? The first step is to sterilize the jars, usually in a pressure cooker, and then to pressure-cook the jam or pickles in the jar. The pressure cooker elevates the boiling point of water, and the higher temperature can kill the endospores. Similarly, medical instruments are "boiled" under high pressure in an autoclave, making them sterile for use in operations. If canned food is not sterilized properly, endospores can survive. One group of endospore-forming bacteria are those that produce a toxin that causes botulism, a food poisoning. Another group produces a toxin that causes tetanus, a serious nerve disease, which is one bacterial infection that does have a vaccine. Puncture wounds, because they admit little air, are prime sources for tetanus infections.

 **Exploration Activities**

1. What is an endospore?

2. Explain how monerans can survive in such varied environments.

Top Shelf Science: Biology MONERA AND PROTOCTISTA

**BACKGROUND**

## Kingdom Protoctista

Protoctists, or *protists,* as they are sometimes called, are distinguishable from the rest of the kingdoms of life in only one regard—they do not belong in any of the other kingdoms. Widely varied, the kingdom is one of exclusion. If an organism is not an obvious moneran, fungus, plant, or animal, it can usually be found in kingdom Protoctista. Beyond that, the organisms that make up the kingdom have very little in common with one another.

Some protoctists are single-celled, while some are multicellular and gigantic. They all require a moist environment—either within water or in moist soil. Beyond the water requirement, protoctists can be found almost everywhere on Earth.

Protoctists are *eukaryotic*—true cells, with full differentiation within the cell, even if they are single-celled. That is, their cells possess a nucleus that contains the DNA of the organism, as well as other cellular structures and organelles. We will look more deeply at the structure and reproduction of the eukaryotic cell in the next section.

Some protoctists are autotrophic—they manufacture their own food supply through photosynthesis. Others are heterotrophic and consume other living things in order to survive. Most are capable of asexual reproduction, but sometimes reproduce sexually when environmental stresses, or the needs of the protoctist, demand it.

There are three basic "types" within the kingdom Protoctista. Until very recently, the several phyla containing the protozoans were thought to be part of the animal kingdom. Protozoans are all single-celled and heterotrophic. They are able to move in animal-like ways by use of **cilia,** tiny hair-like appendages that move the creature through water, or with a whip-like tail known as the flagellum. Some, like the amoeba, send out cytoplasm-containing extensions of the cell membrane. These temporary appendages are known as *pseudopodia.*

> **Widely varied, the kingdom is one of exclusion.**

> **Algae also produce more than half of all oxygen generated by Earth's photosynthesizing organisms.**

Another large group of protoctists are algae. Algae can be either unicellular or multicellular, and are all photosynthetic, like plants. They include dinoflagellates, diatoms, and the ocean and pond algae that are brown, red, and green. Some are plant-like, in that they remain in one location, while many float in the light zones of the ocean and make up the basis for the oceanic food chain. Algae also produce more than half of all oxygen generated by Earth's photosynthesizing organisms. Together with cyanobacteria, they dwarf the amount of oxygen produced by terrestrial plants.

If the protozoans are like animals, and the algae are like plants, the slime molds and water molds are more like fungi. This type of protoctist behaves and reproduces much like a typical fungus. There are several species of slime and water molds.

 **Exploration Activities**

1. Explain why the kingdom Protoctista is a "kingdom of exclusion."

2. What are the three major kinds of protoctists?

3. What distinguishes kingdom Protoctista from kingdom Monera?

## Protozoans

As mentioned, one of the large groups of protoctists are animal-like. Collectively, they are known as *protozoans* (or "first animal" organisms). Protozoans are all single-celled, heterotrophic organisms. They are classified into phyla based on their method of locomotion.

Some protozoans do not move at all—they are *nonmotile*. They are exclusively parasitic, living attached to the intestine or liver, or floating within the bloodstream of an animal, where a ready supply of food can be found. Some of them can cause severe illnesses, such as malaria, giardiasis, and yellow fever. These diseases have killed billions of people worldwide. Some are spread by water, and others have an arthropod *vector*—they are passed along by mosquitoes, ticks, biting flies, and other creatures. In general, the arthropods are not harmed by the protozoans living within their tiny guts and salivary glands, but the insect or tick passes the protozoans into other hosts when it bites another organism. If the organism is susceptible to the illness, over time, as the protozoans reproduce within the host, the organism succumbs to disease. Most of these species are called *sporozoans,* because they can form spores as part of their reproductive cycle.

Some protozoans move by extending a cytoplasm "foot," moving toward it, then sending out the foot again. These are the amoebas, and they are often described as looking like a blob. This is because they have no rigid cell wall—only a membrane, which allows them to change shape at will. You can see an amoeba if you look at any sample of pond water. The "foot" has a special name—*pseudopodia,* and it is really just a part of the membrane, filled with cytoplasm, that seems to ooze in the direction the amoeba wishes to move. The pseudopodia also is instrumental in capturing and absorbing nutrients in the environment. The amoeba uses the pseudopodia to surround a complex organic cell, such as a bacterium, and then draws it into itself, creating a small *vacuole,* or membrane-bound temporary storage chamber, within its "body." Eventually, the nutrients are absorbed for the growth and reproduction of the amoeba.

> Amoebas live in both fresh water and salt water, and even in wet soil.

Amoebas live in both fresh water and salt water, and even in wet soil. Some of them form shells. The two shell-forming varieties of amoeba are foraminiferans, which make a spiral shell out of calcium carbonate and are entirely marine, and radiolarians, which make a very complex shell made of silica and live in fresh and marine environments. They move by extending the pseudopodia through one of the holes in the shell. When times get difficult, they can form a cyst, which, if absorbed by animals, can cause amoebic dysentery. This intestinal illness causes cramps and diarrhea and is associated with drinking water in less than sanitary conditions. In Mexico, it is sometimes called Montezuma's Revenge.

Another phylum of the protozoans includes the flagellates, small organisms that move via one or more whip-like **flagella**. Some are parasitic, and others exist in mutual relationships with hosts, but many more live freely in fresh and salt water.

The most complex of the protozoans are the ciliates, which move by beating small, hair-like cilia. They do this in a coordinated way, which moves the ciliate in the direction it wishes to go.

Ciliates are complex; they have something approaching an almost animal system organization level. They take in food through an oral groove and pass it along to a gullet, where digestive enzymes break it down. The nutrients are spread into the cytoplasm, and wastes are removed through an anal pore. Water enters at all sides through osmosis but is pumped out by *contractile vacuoles* if the salt level within the cell drops too low. Ciliates, although normally asexual, have the capability of sexual reproduction. Sexual reproduction is almost always a response to stress in the environment that requires a more diverse gene pool. Some ciliates, such as the paramecium, have two nuclei. The smaller one controls sexual reproduction, while the larger one controls the other life functions of the organism.

 **Exploration Activities**

1. Describe how an amoeba moves and obtains nutrition.

2. What diseases can be caused by protozoans? Can you think of any way to prevent disease transmission by these parasites?

3. How are flagellates different from ciliates?

## BACKGROUND

# Algae

Algae have one thing in common—they are all capable of photosynthesis and manufacture their own food. They are autotrophs. There are four different kinds of chlorophyll, and depending on the kind or kinds the algae in question makes, its color may be very different from the green color we associate with terrestrial plants.

There are six phyla of algae. Three of them are single-celled: euglenoids, which have both animal and plant tendencies; diatoms, which are unicellular organisms with ornate and beautiful shells; and dinoflagellates, which can move on their own power through flagella. The other three phyla belong to the larger, freshwater and marine algae—green, red, and brown algae.

Euglenoids can ingest food from their surroundings, but they are also capable of making food from sunlight. Along with the dinoflagellates, euglenoids can move via use of flagella.

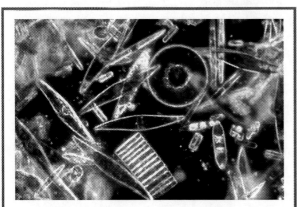

Salt-water diatoms under a microscope

Diatoms are shelled organisms that live both in salt water and in fresh water, as well as in damp soil. They are all members of a phylum called Bacillariophyta. There are two types of diatoms. Pennate diatoms are shaped a little like cylinders and live in fresh or shallow ocean environments. Round, centric diatoms live primarily in ocean water. They are the basis for the oceanic food chain. Diatoms can reproduce asexually, but as they do, their shell splits in half and rebuilds. Thus, the daughter organisms are slightly smaller than the parent was. Eventually, some diatom generation must engage in sexual reproduction, which produces larger offspring who construct an entirely new cell wall.

Dinoflagellates are the last of the unicellular phyla. They come in a great variety of shapes and sizes. Because they possess two flagella, they spend a lot of time spinning slowly through the water. Some species are freshwater, but many dinoflagellates are marine.

Dinoflagellates live mutually with jellyfish, some mollusks, and some corals. Some of them give off an eerie light—they are bioluminescent. Some species can produce deadly toxins. One species produces a nerve toxin that routinely infects mollusks and other sea life and can be deadly to the animals, including humans, that consume them. In warm weather, the number of these organisms can become so large that the sea seems to turn a deep red. Red tide kills many fish and renders shellfish unfit to eat. When red tide is expected or has occurred, shellfish bans are issued.

Red algae has nothing to do with red tide. It is exclusively marine and multicellular. Red algae grows along rocky seashores and in tranquil tropical waters. They attach themselves to rocks by a structure called a *holdfast*. None of the algae possess a root system, such as plants do. There are about 4,000 species of red algae. Some are edible and make up a portion of sushi.

The large kelp forests of California are members of the brown algae phylum. Almost all brown algae live in salt water in cool regions of the world. Of these, kelp is the largest species. Some kelp organisms can measure 100 meters in length! Like red algae, the kelp attaches to rocky shores by use of holdfasts. Many species of brown algae also have air bladders to keep them close enough to the surface of the water to perform photosynthesis. Find a wrack of brown seaweed at the beach. You can usually find one of these air bladders easily. They look like small brown balloons.

Off the California coast, giant kelp forests grow. They are a rich ecosystem, home to a great variety of marine animals, such as fish, abalone and other shellfish, seals and sea lions, and sea otters, which use the kelp forest as a squirrel uses a forest of trees.

Green algae are primarily freshwater, but some live in the oceans, wet soil, and occasionally, the fur of the slow-moving sloths that live in the rainforests. They do not cling to the rocks, preferring instead to float freely in the water. Some species of green algae are single-celled, while others are multicellular. Pond scum, if it is not cyanobacteria, usually turns out to be green algae. Other species are unicellular, but live in large colonies for protection. Green algae have cell walls made of cellulose, like plants, and are believed to be the ancestors of all modern plants.

Large algae engage in both asexual *budding off* and sexual reproduction. Like many animals and plants, algae produce sex cells called gametes. These combine with one another to produce a *sporophyte,* which produces spores and new adult organisms.

> **Some kelp organisms can measure 100 meters in length!**

 **Exploration Activities**

1. What are three exclusively unicellular algae phyla, and how are they unique from one another?

2. Explain why diatoms must occasionally engage in sexual reproduction.

3. Why is kelp an important natural resource in the waters off California's coast?

4. In what unlikely place might you find green algae living?

# Slime Molds and Water Molds

Algae are very much like plants, while protozoans are very much like animals. There are also protoctists that are very much like fungi.

There are three phyla of fungus-like protoctists. Two of those phyla are made up of slime molds, and the other contains water molds and downy mildew.

Slime molds live in cool, damp, shady places, where they subsist on dead or decaying organic material. They can be brilliantly colored—some slime molds are bright yellow, while others are a rich blue or violet. There are two different varieties of slime molds—plasmodial slime mold and cellular slime mold.

Plasmodial slime molds form a *plasmodium*—a mass of cytoplasm without cell walls or membranes but with multiple nuclei. There is always more than one organism in these large bodies—sometimes there are thousands. A plasmodium may reach a size of 1 meter in diameter. The various members of the plasmodium form spores from sexual reproduction, which are dispersed by wind. A cellular slime mold exists as an individual for its entire life. Even when congregating for mating, it retains its own cells, each with its own cell membrane.

Water molds live in water or moist places. Some parasitize plants, while others feed on dead organisms, such as fish. They are fuzzy, white growths that are easiest to see on decaying matter. Like the mold on a rotting piece of fruit, they form long, fuzzy, white threads and digest the nutrients within whatever they are eating.

Water molds and downy mildews are responsible for serious plant diseases around the world, some of which have resulted in famine and mass starvation.

A water mold caused the Great Potato Famine (1845–1849) in Ireland. The entire potato crop in that country was destroyed over several growing seasons. Heavy rains kept the mold alive. The famine killed thousands and displaced many Irish farmers, some of whom emigrated to America.

> Slime molds live in cool, damp, shady places, where they subsist on dead or decaying organic material.

 **Exploration Activities**

1. How do plasmodial slime molds reproduce?

2. What is the difference between cellular slime molds and plasmodial slime molds?

3. How have water molds negatively affected human beings?

# Student Lab: Growing Water Molds

In this experiment, we will grow a water mold. The spores that will form the mold already exist in the water or in the hamburger meat; however, both are necessary to begin the process of growing the mold.

## Materials

- Small bit of uncooked hamburger or other meat
- Plastic container full of water
- Microscope, slide, and cover slip
- Iodine stain

## Procedure

1. Place a small amount of hamburger in the water and leave it for a few days in a sunny location.

2. Notice that a transparent mass begins to grow on the hamburger after one or two days. Collect some and prepare a slide by staining the mold.

3. Observe under low power. You will see that the water mold has a cell structure like that of amoebas.

## Conclusion

Where did the water mold come from? What is its role in aquatic ecosystems?

# Eukaryotic Cell Structure

 BACKGROUND

**The nucleus is the defining feature of the eukaryote.**

All living things are made up of one or more cells. As we have seen, some cells are extremely simple—the prokaryotes, or first cells. After the moneran prokaryotes, the next evolutionary step was eukaryotic cells. Eukaryote means "true cell," and the cells have well-defined nuclei, as well as many other **organelles**, that are separated from the cytoplasm by membranes. All living organisms beyond the kingdom Monera—including Protoctista, Fungi, Plantae, and Animalia—have at their base the eukaryotic cell structure.

Except for their lack of a cell wall and chloroplasts, animal cells look like every other eukaryotic cell. Here, then, are the basic parts of the typical eukaryotic cell:

- **Nucleus** The nucleus is the defining feature of the eukaryote. It contains the entire set of the organism's DNA in long, tangled strings called **chromatin**. Also within the nucleus is the **nucleolus**, where tiny cell particles involved in protein synthesis are produced. These particles, called **ribosomes**, eventually move throughout the cell, and are the locations where the cell produces enzymes and other proteins to carry out the message of the DNA.

- **Nuclear Envelope** The nucleus is surrounded by a *nuclear envelope* with pores that allow the transport of *nucleotides*—sugars used by the nucleus—and proteins into the nucleus, and ribosomes out of the nucleus.

- **Cytoplasm** As previously mentioned, cytoplasm is the gel-like liquid within the cell. In a typical animal cell, the cytoplasm is about half the volume of the total cell. This liquid cushions the organelles within the cell and keeps internal structures in place.

- **Endoplasmic Reticulum** This is a folded membrane that provides a large surface area on which chemical reactions take place for the cell. The membranes also contain the enzymes, and serve as the site, for the cell's *lipid* (fat) synthesis. Some of the endoplasmic reticulum is coated with bumpy ribosomes, while some is smooth.

- **Golgi Apparatus** The Golgi apparatus is a series of stacked membrane sacs that stores synthesized proteins and lipids from the ribosomes and endoplasmic reticulum, and delivers them to the cell plasma membrane and all the other organelles. While in the Golgi apparatus, the proteins and lipids are modified and "repackaged" into a form the other organelles can use.

- **Vacuoles** As described in the section on protozoans, vacuoles are storage chambers within the cell. Food, water, enzymes, and waste products are all stored temporarily by cells within vacuoles.

- **Lysosomes** Lysosomes are organelles that contain digestive enzymes and digest worn-out cell parts, food particles, and invading bacteria or viruses. The membrane of this particular organelle is a one-way street. The corrosive digestive enzymes cannot escape to cause damage to the cell.

- **Plasma Membrane** Also known as the cell membrane, this is the outer limiting membrane of an animal cell, and it regulates transportation into and out of the cell.

- **Mitochondria** These are where cell respiration takes place. The mitochondria also produce **ATP,** or adenosine triphosphate, which are energy-storing molecules. Mitochondria have their own DNA, which is linked directly from the female line of succession and is different from the DNA in the nucleus. To distinguish between normal DNA and mitochondrial DNA, mitochondrial DNA is written as mDNA.

- **Cell Wall** Plants, some protoctists, and many fungi contain a cell wall. This is a rigid outer shell, beyond the plasma membrane.

- **Chloroplasts** Eukaryotes that are autotrophic also have small organelles called chloroplasts that store chlorophyll.

- **Cytoskeleton** The cytoskeleton is a network of filaments that give shape and structure to the cell, allow for the transport of materials within the cell, and allow for contraction and motion of the cell.

> **Plants, some protoctists, and many fungi contain a cell wall.**

 **Exploration Activity**

Create a model of a basic animal or plant cell in the medium of your choice. Be sure to label all internal structures.

## Heredity and Genetics

We receive many of the things that make us unique—our eye color, our hair color, our height, and many other characteristics—from both of our parents. Because our species reproduces sexually, we have a mix of characteristics that are inherited from one or the other parent. Not only are our *genetic codes* mixed up like this, so are the genetic codes of almost all organisms on the planet. The only organisms that do not receive information from two parents are those with only one parent—monera, some protoctists, and simple plants and animals when they bud off rather than share DNA.

It may surprise you to learn that humans did not always understand this. Humans have believed various things about reproduction over the years—that males, or females alone, gave the developing embryo and fetus its characteristics, or that the fetus could be changed *in utero* by shock, diet, or other events during the pregnancy. Only relatively recently did humans understand that the human male sperm, with its potential Y chromosome (females carry only X chromosomes), determines the gender of the offspring. Modern laws of heredity—of humans, other animals, plants, and all other sexual organisms—were developed based on the research of one man, Gregor Mendel, who was an Austrian monk, mathematician, and gardener during the nineteenth century.

Mendel focused on the heredity of pea plants in his garden. Peas grow rapidly, flower early, and produce many seeds, which allowed Mendel to produce several generations of peas every year. He pollinated some plants by others (*cross-pollination*), and allowed others to *self-pollinate*. He then recorded certain characteristics—size of stems, texture and color of the peas, shape of the peapods, and position of the flower—in the young plants. Usually, traits in one or the other of the parents were seen in the young plants; however, this was not always the case. Mendel learned that traits that were apparent in the parents were not always expressed in the offspring. Likewise, traits might surface in the offspring that had not been seen in the parents.

> Because our species reproduces sexually, we have a mix of characteristics that are inherited from one or the other parent.

> Today, we know that the chromosome pairs in the parents' cells separate to form gametes, or reproductive cells, and join to form a zygote, or fertilized gamete.

Mendel called the traits seen in the parent that were transmitted to the offspring *dominant* traits. The contrasting traits of the other parent did not seem to show up in the offspring. He called the trait that seemed to disappear *recessive*. For instance, he cross-pollinated a plant that had smooth green peas with one that had wrinkled yellow peas. The offspring had smooth green peas, so Mendel concluded that smooth and green were dominant traits, and yellow and wrinkled were recessive traits. However, if Mendel continued to cross-pollinate successive generations of this family of plants, eventually, he would have produced plants with yellow, wrinkled peas.

Today, we know that the chromosome pairs in the parents' cells separate to form *gametes,* or reproductive cells, and join to form a **zygote**, or fertilized gamete. The characteristics that make up the gene pair have several possible outcomes. The dominant *allele,* or gene for a particular trait (for green) can be paired with the other dominant allele, and green seeds result in the offspring. Or the dominant allele can be paired with the recessive allele (for yellow). If this happens, the offspring will still be green, because the dominant allele is expressed when paired with the recessive one. But if both parents have a recessive yellow allele, at some point, the two recessive alleles will pair up, and the result will be a plant with yellow seeds.

Here's a diagram of the possible outcomes:

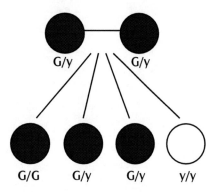

A statement can be made about the probability of seeing a recessive trait surface, based on known characteristics of the parents. Mendel was able to do this because his parent plants were *purebred*—they all shared the same characteristics without deviation for several generations. On the other hand, his offspring plants were *hybrid*—their parents had differing traits.

When genes in an allele pair are identical (GG or yy), they are said to be *homozygous*. If the genes are not identical (G/y), they are *heterozygous*.

> **Mendel's ideas go far beyond the vegetable garden.**

Mendel's ideas go far beyond the vegetable garden. One easy way to see Mendel's principles in action is to look at the eye color of your family members. Brown eyes are dominant in humans, and light (blue, green, hazel) eyes are recessive. If your parents have brown eyes, they can have light-eyed children, because they each might have the recessive gene for light eyes. However, the more likely situation is that most, if not all, of their children will have brown eyes. If your parents have light eyes, however, all the children will also have light eyes, because the recessive gene is expressed in the parents, so neither of them has the dominant gene to contribute to the mix. If one parent has dark eyes and one has light eyes, there will most likely be a mixture among the children.

 **Exploration Activity**

Design an experiment with plants that shows dominant and recessive characteristics in the offspring of two parents with differing characteristics. The characteristics can be size, color, or any other obvious physical difference. Run the experiment and determine which of the parents' characteristics were dominant or recessive, based on the number of offspring with each characteristic. (Example: Grow two colors of petunias, cross-pollinate, and see what color flowers grow from the seeds that result.)

## Ploidy, Meiosis, and Mitosis

Under ordinary circumstances, most sexually reproducing cells have two sets of chromosomes within their nucleus. One set comes from one parent of the organism the cell belongs to, while the other set comes from the other parent. If the organism has reproduced asexually, both sets come from the same parent.

Any cell with two sets of chromosomes is called **diploid**. Most eukaryotic cells—virtually all animals, most plants, and most fungi—are diploid, although a rare few have only one set of chromosomes (**haploid**), and even fewer have more than two sets (*polyploid*).

**Ploidy** refers to the number of chromosomes within the nucleus of the cell. For most organisms, the ploidy only changes in selected cells during sexual reproduction.

Cells grow and divide almost constantly. When you cut yourself, and cells grow and divide to heal the wound, or when a small child grows an inch, the population of cells has increased. This process of cell division is called **mitosis**. There are four phases of mitosis. The first, the *prophase,* is characterized by the DNA chromatin coiling up into visible chromosomes. Each duplicated chromosome is made up of two halves, called **sister chromatids**. They are exact copies of each other and are attached to each other by a membrane called the **centromere**. As the prophase continues, the nucleus seems to disappear. Two important structures—the **centrioles**, which are dark, cylindrical structures—begin to move to opposite ends of the cell. Between them forms the **spindle**, a cage-like object.

In the second phase, *metaphase,* the doubled chromosomes become attached to the spindle. They are pulled by the **spindle fibers** until they are neatly separated on the midline of the spindle. One sister chromatid of each chromosome is attached to one spindle fiber, while the other is attached to the other spindle fiber.

In the third phase, *anaphase,* the separation of the sister chromatids begins. The centromere splits, allowing the two new chromosomes to separate fully. In the final phase, *telophase,* the chromatids have moved to the opposite ends of the cell, following the centrioles.

> Any cell with two sets of chromosomes is called diploid.

The spindle breaks down, and a new nucleus forms around each of the new sets of chromosomes. Finally, the cytoplasm divides, and two new cells are formed.

Between the reproduction stages, the cell is an extremely busy place, duplicating DNA, making ATP, repairing its membranes, producing new organelles, and so forth. This time is known as *interphase.*

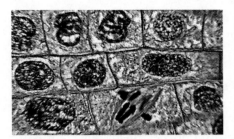

Onion cells undergoing mitosis. This slide shows several stages of the process.

Mitosis guarantees genetic continuity. The two daughter cells have the same genetic code and fulfill the same functions as the parent cell did.

In organisms that reproduce sexually, as Mendel's pea plants did, new organisms are not formed by mitosis, but by **meiosis.** Meiosis is the process by which sex cells are formed. Unlike mitosis, which produces the same number of chromosomes in each new cell as the parent, meiosis reduces the number of chromosomes in each sex cell produced by half.

In the first meiotic prophase, each pair of chromosomes comes together to form a four-part structure called a *tetrad,* which consists of two chromosomes, each made up of two chromatids. They are packed so tightly together that non-sister chromatids sometimes exchange genetic material in a process called *crossing over.*

In the first metaphase, the tetrads line up on the midline of the spindle, instead of the chromosome pairs lining up on the spindle independent of each other. During the first anaphase, the chromosome pairs separate and move to opposite ends of the cell. The centromeres, however, do not split, as they do in mitosis. This critical difference ensures that each cell will receive only one chromosome, instead of two, from each pair. Finally, in the first telophase, the spindle is broken down, the chromosomes uncoil, and the cytoplasm divides to create two new cells. But meiosis isn't over yet.

> Whereas mitosis provides for genetic stability, meiosis provides for genetic variation.

Instead of then replicating the DNA, another period of meiosis begins. During the second prophase, the process of spindle formation takes place. Again in the second metaphase, the lining up at the center of the cell takes place. This time, however, in the second anaphase, the centromeres *do* separate and move to opposite sides of the cell. Finally, the nuclei reform. At the end of the process, in the second telophase, four cells have been formed from the original diploid cell. These cells are no longer diploid—they are haploid. Each cell has only half of the information the parent cell contained—one chromosome from each of the pairs.

The haploid cells will go on to become **gametes,** or sex cells. They will combine with the opposite sex cell of another organism in order to create a new, functionally different organism.

Whereas mitosis provides for genetic stability, meiosis provides for genetic variation. The shuffling of the chromosomes provides a hedge against future illness or other kinds of problems in the environment. One set of genes might be susceptible to disease or environmental stress. By having many individuals with many different characteristics in the gene pool, the chance of species survival improves.

 **Exploration Activities**

1. What is ploidy? Are sex cells haploid or diploid?

2. Describe the process of mitosis.

3. What is the result of meiosis? Why are sex cells important?

# Student Lab: Observing Cell Division in Onions

In this lab, we will be watching the very process of cell division in onion root tips. Over time, we will be able to see all the main steps in cell division.

## Materials

- Microscope, slide, and cover slip
- Onion root tips
- Iodine stain

## Procedure

1. Prepare a slide of a longitudinal section of an onion root tip and stain with a drop of iodine.

2. Examine the slide using low power.

3. Locate the region most likely to have dividing cells, just behind the root cap. At the tip of the root is the root cap, which protects the tender root tip as it grows through the soil. Just behind the root cap is the zone of greatest cell division.

4. Switch to intermediate power, focus, then switch to high power.

5. Survey the zone of cell division and locate the stages of the cell cycle—interphase, prophase, metaphase, anaphase, and telophase.

6. Locate the nucleus, chromosomes, spindle, and sister chromatids.

7. See if you can spot and name the other organelles in a mature cell.

## Conclusion

Do all the cells in a particular region reproduce simultaneously? Why or why not?

# Kingdom Fungi

Fungi are a group of heterotrophic, usually multicellular organisms that are instrumental in the decomposition process. Members of this kingdom, which is the smallest of the five kingdoms of life, can be found almost everywhere on Earth. The kingdom includes mushrooms, truffles, morels, ringworm, molds, rusts, and yeasts.

When we look at a fungus, we are typically seeing only the tip of the iceberg, so to speak. Most of the fungus is hidden from view, deep in the leaf litter the fungus is busily decomposing, or even under the ground. To give an example of the kind of scale a fungus can encompass, consider the individual mushroom fungus that lies under the Upper Peninsula of Michigan. The single individual of the species *Armillaria bulbosa* covers a total of 38 acres and is estimated to weigh about 100 tons. The fungus is ancient—at least 1,500 years old. Here and there, the fungus erupts above the ground in small mushroom circles. But below the ground, there is a long network of bundles of filaments called mycelia, weaving around the trees in the forest and going virtually unnoticed until mushrooms sprout. The bulk of the organism is hidden from view, quietly cleaning up the forest floor.

Fungi usually recycle nutrients from dead organic material. In so doing, they are an important part of the ecosystem. The nutrients are returned to the soil or water, where they are made available for plants and other photosynthetic organisms to use to make food, starting the food chain anew.

While fungi are mostly decomposers, occasionally they parasitize living organisms. Many minor, irritating infections, such as athlete's foot, as well as more serious diseases in plants and animals, are caused by fungi. Plant rusts that develop from fungi often wipe out entire harvests of grain or render it dangerous to eat. Rust and black spot on roses are caused by fungal infections. Some fungi exist in mutual relationships as well, as we saw with the lichen experiment in the first section of this book.

> Members of this kingdom, which is the smallest of the five kingdoms of life, can be found almost everywhere on Earth.

> Some fungi are used to make antibiotics and other vital medications.

On the other hand, some fungi are instrumental in food preparation and fermentation of beer. Some fungi are used to make antibiotics and other vital medications. And of course, some fungi are edible, and quite tasty, such as many species of mushrooms and truffles. (Poisonous mushrooms kill quickly and painfully, and often look like the edible variety. Never collect wild mushrooms unless you are an expert! Each year, people die or need emergency liver transplants from eating toxic mushrooms. Good, fresh mushrooms can almost always be purchased at grocery stores, and they are guaranteed safe.)

Historically, fungi were classified as plants, because they possess a cell wall and root-like structures. But fungi do not perform photosynthesis, and they are not autotrophs. Their cell walls are made of a very different substance from the cell walls of plants, as well. *Chitin* is the same complex carbohydrate that forms our fingernails and hair, and the exoskeletons of arthropods. Plant cell walls are usually made of cellulose. As scientists learned about these differences, and about the life cycle and style of fungi, they were reclassified into their own distinct kingdom.

While there are a few single-celled fungal organisms, such as yeasts, most fungi are large and multicellular. The basic structural unit of fungi are the hyphae. Hyphae are long, threadlike filaments that develop directly from **spores**, which are the reproductive cells of the fungus. The hyphae branch out at one end to interlock with other hyphae, forming a mass called a **mycelium**. Inside individual hyphae, there are usually divisions called septa. Each *septum* has small pores, through which cytoplasm and organelles flow freely from one cell to the next. This is the method by which fungi move nutrients throughout the organism.

Food is digested outside of the fungal cells and then absorbed. The fungus releases powerful digestive enzymes into the object about to be consumed. When the large organic molecules break down, the smaller ones **diffuse** (move from an area of larger concentration into the area of smaller concentration) into the fungus. They are moved along through the septa to where they are needed for growth, cellular repair, or preparation for reproduction.

Fungi are capable of sexual and asexual reproduction. Asexual reproduction, in the form of fragmentation, occurs when a piece of a large fungus breaks off and begins to grow a new mycelium. This occurs with regularity as people garden or dig wells, or root up tree stumps. Budding off is a form of mitosis. When a fungus buds, two new nuclei are created. A fungus that generally reproduces by budding is the single-celled yeast fungus.

Fungi can also reproduce both sexually and asexually through spores. Spores are reproductive cells that germinate, not unlike seeds, into new organisms. A fungus produces spores, and when the spores are released, they are carried via wind, water, or animal vector, sometimes long distances. When the spore lands in a place that is optimal for growth, it germinates—a hypha appears and begins to grow and branch to form a new mycelium.

Eventually, some of the hyphae grow upward and produce a spore-bearing structure called a *sporangium*. These objects vary widely with the type of fungus. They are small sacs or cases that often appear on whatever the fruiting body is—the mushroom cap, for instance, or the black spots in bread mold. The spores can be produced by mitosis or meiosis. Eventually the sac ruptures and releases the spores. Spores are very lightweight and can easily be swept along by wind or small insects.

Fungi are classified by how they form sporangia during reproduction.

> **Fungi are classified by how they form sporangia during reproduction.**

 **Exploration Activities**

1. Explain why we can only see a tiny percentage of a fungus organism.

2. What vital role do fungi play in the ecosystem?

3. What is the basic structural unit of a fungus? What does that unit eventually form?

4. List the three methods of reproduction among fungi.

5. What is a sporangium? How is it formed, and what is its role?

## Zygospores and Sac Fungi

Many have had experience with *Rhizopus stolonifer,* a black bread mold. It, along with all its close relatives, are within the phylum Zygomycota. Most of these species are decomposers—there are few parasites or mutual organisms among them—and they produce small black spores.

When a spore settles on a piece of bread, or any other good location, the hyphae begin to grow. Some grow along the bread, and others grow down into the bread. The ones that grow along the bread, horizontally, are called the *stolons,* and they can be seen by the sandwich maker. The ones that grow down within the bread are called the *rhizoids*. It is near the rhizoids that the digestion of the bread takes place.

During reproduction, some hyphae grow upward and develop little black sporangia at their tips. This form of reproduction is asexual. The organism can also reproduce sexually, if two different organisms are present on the same piece of bread, and the tips of two different hyphae fuse. If they do this, the spores that form have thicker walls and can withstand unfavorable conditions, such as drying out. A zygospore may lie dormant for years and withstand dry conditions, cold, and heat. When the conditions improve, the spore germinates and begins its life cycle again.

The next phylum, Ascomycota, contains the largest number of species. They are sometimes called the sac fungi, because they form small sac-like structures during reproduction. Some are edible, such as morels and truffles; some are very useful, such as yeasts; and some are very beautiful. However, many of them carry plant and animal disease, as well. Dutch elm disease wiped out most of the native elms in North America, ergot on rye plants led to mass poisonings during the Middle Ages, and apple scab is a problem apple growers face even today.

Spores are produced inside a small sac called an *ascus,* and are called *ascospores*. Like zygospores, they can reproduce sexually or asexually.

> A zygospore may lie dormant for years and withstand dry conditions, cold, and heat.

Top Shelf Science: Biology  FUNGI AND PLANTS **Zygospores and Sac Fungi**

 **Exploration Activities**

1. Draw a life cycle of a zygospore-bearing fungus in its asexual reproduction phase.

2. How have sac fungi helped humanity? How have they been harmful?

**BACKGROUND**

# Club Fungi and the Deuteromycotes

Probably the most familiar fungi belong to the phylum Basidiomycota, the Club Fungi, which includes mushrooms, puffballs, and stinkhorns, as well as the smut fungi, which are serious agricultural pests.

The spores of the phylum Basidiomycota are produced in club-shaped hyphae called *basidia*. The reproductive cycle for mushrooms and other Basidiomycotae are somewhat complicated. The mushroom typically consists of a large thallus, or stalk-like object, sometimes called a *stipe,* which supports a cap. Under the cap are *gills,* a series of membranes. The photo below, of the orange mycena group of mushrooms, shows the gills under the mushroom caps very clearly. Along the gills, the spores are contained in small club-shaped basidia. As the cap decays, wind, rain, or animal fur carries away the spores. When the spores land in a suitable environment, they germinate into hyphae, which grow down into the soil.

> **The spores of the phylum Basidiomycota are produced in club-shaped hyphae called basidia.**

Orange mycena group

Under the ground, a mycelium forms, with only one set of chromosomes. It is a haploid organism. The mycelium must mate with another organism to produce a diploid offspring.

The two haploid cells fuse and form *buttons*—compact masses of hyphae, just below the surface of the soil. The buttons develop into mushrooms. Inside each basidium, the club-shaped spore-producing cell, the two haploid nuclei come together to form a diploid cell. Meiosis then occurs, and new nuclei are produced that become part of the spores.

Unlike the complex reproductive process of the Basidiomycotae, the Deuteromycotae have no sexual stage in their life cycle. One very important member of this phylum is *Penicillium notatum,* the fungi that produces the antibiotic credited with saving millions of lives. Other members of the phylum play important roles in making cheeses, soy sauce, and citric acid.

## Exploration Activities

1. Describe the reproductive cycle of a mushroom.

2. How are members of the phylum Deuteromycota useful to humans? What sets them apart from the other fungi?

**BACKGROUND**

> The sun provides all necessary energy to the ecosystem by giving producers what they need to perform the task of photosynthesis.

# Fungi in the Nutrient Cycles

Food chains show how energy and nutrients move from producers through first-, second-, and third-order consumers of an ecosystem. The sun provides all necessary energy to the ecosystem by giving producers what they need to perform the task of photosynthesis. A portion of that energy passes up the food chain until the last animal in the chain dies. At that point, the energy from the sun is lost. Since the sun keeps shining, this does not cause a problem for living things. The nutrients that are not solar in origin must be recycled. They are returned to the soil or air for a new generation of plants to use.

The matter that makes up the cells of all living organisms alive today—carbon, nitrogen, and the oxygen and hydrogen that make up water—are the same atoms that have been on Earth since the beginning. They are constantly recycled. Water is cycled through *transpiration* from trees and other plants, and evaporation from lakes and the oceans. It is then returned to lakes, oceans, and groundwater tables through *precipitation,* and becomes available for organisms to use.

Carbon and nitrogen, however, require that the bodies of once living things be broken down to their molecular level, and most importantly, that the form of the nutrient in the soil or air is one that the producers can use.

Carbon is found in the atmosphere as $CO_2$. Carbon dioxide plays an important role in the nutrient chain, because it is necessary for photosynthesis. All living things on Earth are carbon-based. Our bodies, the bodies of sea stars, and the molecules of plants are all made up of carbon molecules. We obtain the carbon we need to repair our bodies and reproduce from the food we eat. When organisms die, decomposers such as fungi and bacteria break the solid carbon down to atmospheric molecules, so that plants can take it back up again during photosynthesis.

> Fungi, together with bacteria, break down formerly living matter and close the nitrogen cycle, making life possible on Earth.

Atmospheric nitrogen makes up more than three quarters of the air. However, living things cannot use nitrogen in that form. Some nitrogen can be made usable if lightning breaks molecules apart in the atmosphere, but this amounts to only a very small fraction of the needed amount of nitrates. Animal urine is also directly usable, but again, this is a small amount of what plants and other photosynthetic organisms require. Some bacteria, the proteobacteria, also fix nitrogen in the soil directly, but the main source of nitrogen usable by plants is ammonia, which is a by-product of decomposition. Fungi, together with bacteria, break down formerly living matter and close the nitrogen cycle, making life possible on Earth.

 **Exploration Activities**

1. Describe the role of fungi in the carbon cycle.

2. Describe the role of fungi in the nitrogen cycle.

3. Why is recycling nutrients necessary for living things to continue to survive on Earth?

# Student Lab: Bread Mold

As with the water molds, the spores that make up the bread molds are in the environment. In this experiment, we will observe, over time, the growth of common bread molds and watch their reproduction. We may also see bacterial growth.

## Materials

- Bread (preferably a kind with no preservatives, as those inhibit the growth of molds)
- Substrate material (such as soil)
- Sandwich bags with a tie or zipper lock
- Marker to label the bags
- Damp paper towels

## Procedure

1. Label the bags with a number or name so you can tell them apart. In some, place a damp paper towel and a slice of bread; in others, place only a slice of bread.

2. Sprinkle a little substrate material, such as soil, leaf litter, or some other substance likely to harbor mold spores, into the bags and close the bags tightly.

3. Check the bags daily for fungal growth. Fungi should be visible in three to five days. Bacterial colonies may also grow. Fungi will appear to be green, white, yellow, or black and fuzzy. Bacteria will appear slimy and may also be of various colors.

4. Record the number, color, and size of the fungal colonies. The black bread mold may completely cover the bread in a short period of time.

## Conclusions

1. What kinds of mold or bacterial growth did you see?

2. Was the growth more pronounced if a wet towel was used, or if a dry slice of bread was used? Why?

3. What role did the substrate material play in this experiment? How might your findings have been different if you used sand or living leaves?

**BACKGROUND**

# Kingdom Plantae

Plants are all multicellular, autotrophic organisms with eukaryotic cells. They have cell walls, generally made of cellulose, and a waterproof coating called a *cuticle* on leaves, stems, branches, fruits, and seeds.

Aside from these few similarities, plants are remarkably diverse. They can live in the driest deserts, in lakes and ponds, on cliff sides, and at the very edge of the sea. The first living things that return after a fire in a forest are simple plants. The first living things that made the transition from sea to land were plants. They can survive the coldest winters and return anew in the spring. The plant kingdom contains the tallest and oldest living things on Earth.

The first plants made the transition from ocean to land about 500 million years ago. These plants were descended from green algae, which also have cell walls made of cellulose and produce the same type of chlorophyll that plants produce. The first terrestrial plants, like the green algae, were very simple. They most likely looked something like modern-day mosses, and built up the soil enough so that other plants, with slightly deeper root systems, could take hold and flourish. The first plant fossil is more than 400 million years old and is a member of the phylum Psilophyta, which contains modern relatives such as the whisk fern. These plants were made up of stems without leaves and tiny root hairs that pulled moisture and nutrients from the ground.

Life on land had great advantages for plants that, at least for a while, were provided with an environment without many competitors for resources. However, in any terrestrial environment, there are challenges. The first plants lived in swampy, lowland areas where water was fairly easy to come by in the upper layers of the soil. Plants did not need to grow very tall to obtain sunlight, nor did they need long root systems, or even vascular systems to move nutrients around. As the available lowland areas began to fill, plants began to increase in size. This created the need for a more complex system of roots, as well as a branching system containing

> The first living things that made the transition from sea to land were plants.

leaves to capture sunlight. As the plants got taller, it was also necessary for roots to grow deeper in the ground.

With these adaptations already in place, some plants found it possible to live in slightly drier locations. Their longer root systems could tap deeper water sources, and their vascular systems could transport water up to the tops of the plants, nourishing the leaves with water and nutrients from the soil, and sending the simple sugars produced back to the roots for storage. Plants began to produce gametes that were covered with a waterproof coating—spores, cones, fruits, and seeds of all types.

All plants have adaptations that minimize water loss to some degree. Leaves, stems, and fruits are all coated with a waxy substance, made of lipids, called the *cuticle*. Since water does not dissolve lipids, the cuticle keeps the leaves from drying out, while maintaining a wide area for photosynthesis to occur. Under each leaf, there are openings called *stomata*. The pores of the stomata remain open during the day while photosynthesis is taking place, taking in carbon dioxide and releasing oxygen and water. At night, the stomata close up, preventing additional water loss. Typically, a plant may lose up to 90% of the water it pulls up from the soil through the stomata. The release of water from the leaves of plants is called *transpiration* and is an important part of the water cycle.

Some plants live in regions that are so hot and so dry that any loss of water due to transpiration is a bad idea. Some maintain leaves only briefly, performing enough photosynthesis in a day or two to maintain the plant for a year. Some gave up growing leaves altogether and began to photosynthesize through branches or even stems. The waxy cuticle prevents dehydration by slowing the evaporation of water from inside the plants.

Most plants depend on the soil for their source for water, although some survive by collecting rainwater, or living so close to the water's edge (or even within the water itself) that they can tap into ponds and streams. More complex plants have root systems, while simpler ones have tiny root-like objects called *rhizoids*. Most plants also have a *vascular* system for transporting water, nutrients, and the simple sugars made in photosynthesis to where they are needed or are stored. The main parts of the vascular system are the *xylem*, which is tissue made from a series of dead, tubular cells that

> **More complex plants have root systems, while simpler ones have tiny root-like objects called rhizoids.**

> **Mosses do not have a vascular system.**

transports water from the roots up into the plant, and the *phloem*, tissue made from still living tubular cells, which transports sugars from the leaves to all other parts of the plant. Mosses, and a few other small plants, do not have a vascular system. They transport water and nutrients from cell to cell through the process of osmosis. Plants like these are usually close to the surface of water sources.

Plants vary in their reproductive strategies, as we will read when we look at the various plant phyla. But all plants exhibit two *alternating generations*—a generation that produces gametes, or haploid organisms, and a generation that produces spores, or diploid organisms. The generation that produces the gamete is called the gametophyte, and the generation that produces the spores is called the sporophyte.

 **Exploration Activities**

1. Name the features that define the plant kingdom.

2. What is the importance of the cuticle?

3. From what protoctist organisms are plants descended? How do we know?

4. Name some adaptations that allow plants to live in drier climates.

5. What is the system that most, but not all, plants have?

### BACKGROUND

# Photosynthesis

Green plants and other photosynthetic organisms, such as cyanobacteria and algae, convert solar energy into chemical energy by the process of **photosynthesis**. Recall that such organisms are autotrophs. All other species on the planet are dependent upon this solar energy, transformed to chemical energy, for their survival. They obtain this energy by eating plants or by eating animals that have eaten plants. Such species are heterotrophs.

During photosynthesis, plants use the raw materials carbon dioxide ($CO_2$) and water ($H_2O$) to produce food in the form of energy-rich compounds, mainly glucose. A by-product of the process is oxygen.

A substance in photosynthetic organisms called **chlorophyll** is necessary for photosynthesis. Chlorophyll is a pigment, usually green, that gives photosynthetic organisms their characteristic color. Other colors of chlorophyll also exist, as we see in algae, which can be brown, golden, or red, and in cyanobacteria, which are a shade of bluish green.

Here is the equation for the process of photosynthesis. Don't panic! We will break it down.

$$6CO_2 + 6H_2O \longrightarrow C_6H_{12}O_6 + 6O_2$$

The equation demonstrates that for every six molecules of carbon dioxide and every six molecules of water, a photosynthesizer produces six molecules of oxygen (the waste product) and one molecule of glucose, a simple sugar. Actually, the process is more complex. This equation shows the initial substances and the end products. There are several different chemical reactions that take place before the end products.

Some wavelengths of light are absorbed by plants, and some are reflected. Green chlorophyll absorbs mostly red and blue wavelengths and reflects mostly green, which is why plants appear to be green. The absorbance is different for red and brown algae, and for cyanobacteria. In most photosynthetic organisms, the chlorophyll is stored in special organelles called chloroplasts. This is not true in cyanobacteria, which have no organelles. In each chloroplast are

> Green plants and other photosynthetic organisms, such as cyanobacteria and algae, convert solar energy into chemical energy by the process of photosynthesis.

> **The amount of sunlight, water, and carbon dioxide in the air and the temperature all affect the rate of photosynthesis.**

tiny structures called *grana,* where the energy is actually absorbed and released.

Also during photosynthesis, you may have noticed that no $H_2O$ emerges at the other end of the reaction. This means that water molecules have broken down. The energy for this task comes from the grana.

The amount of sunlight, water, and carbon dioxide in the air and the temperature all affect the rate of photosynthesis. The temperature and sunlight requirements are why we do not see many plants growing in the winter; plants conserve their energy until there are many sunny days and the temperature is warm enough for growth.

 **Exploration Activities**

1. Describe the process of photosynthesis.

2. What is the simple sugar that is produced during photosynthesis?

3. Why do plants look green?

4. What are chlorophyll and chloroplasts?

## Spore-Bearing Plants

There are five phyla of spore-bearing plants. One is nonvascular, while the others have some form of veins and transportation system.

The bryophytes, the most simple, are the only plants that are nonvascular. They include mosses and liverworts, and grow in sheltered stream valleys and near ponds, where water is plentiful. All of their life functions, including photosynthesis and reproduction, must take place near water. Because bryophytes do not have veins, they must pass nutrients slowly from cell to cell through osmosis. The lack of vascular tissue limits their size. All bryophytes are small organisms. Despite the limitation that a lack of a vascular system causes, these plants are successful in their specialized habitats. Mosses, indeed, are often the first organisms to colonize locations that have been devastated by fire or other natural disaster. Mosses were the first living things to return to Mount Saint Helens, for instance, in the wake of the destructive volcanic eruption of 1980.

Like all plants, the life cycle of the bryophyte includes an *alternation of generations*. Many bryophytes rarely reproduce sexually. Budding off is the preferred method of liverworts and hornworts. However, if necessary, all bryophytes can reproduce sexually. In mosses, the haploid spore germinates to form a structure called a *protonema*—a small, green filament of cells that produces a gametophyte that has both male and female reproductive structures. Liverworts and hornworts do not form the protonema; instead, the spore germinates directly into the plant body. In either case, the gametophytes produce the male structure, the *antheridium,* in which sperm are produced, and the female *archegonium,* where ova are produced. Sperm are released and swim toward the archegonium, where they fertilize the ova to create the sporophyte generation.

Psilophytes consist of thin, green, leafless stems. The stems are covered with small, leaf-like scales. Most of the species of psilophytes, which include the whisk fern, are confined to tropical or subtropical regions. They produce spores at the tips of their stems.

> Like all plants, the life cycle of the bryophyte includes an alternation of generations.

Lycophyta, also known as club mosses, although they are only distantly related to the mosses, are simple vascular plants that are adapted primarily to moist environments. Modern species are very small, but in the distant past, Lycophyta included very tall organisms—some 30 meters high—during the Carboniferous period of the Paleozoic era. In the lycophyte reproductive cycle, the spore germinates to create a gametophyte structure called a *prothallus*. The prothallus is either male or female and produces only sperm or ova. The sperm must swim through a film of water to reach the ova of a separate plant. The sporophyte grows from the fertilized zygote.

Sphenophyta are the horsetails. Like other spore-bearing plants, horsetails prefer moist environments. They can typically be found along streambeds. Like the Lycophyta, the horsetails once grew to tree height. Horsetails have hollow, jointed stems with scale-like leaves, giving the whole plant the appearance of the tail of a horse. The reproductive cycle of the sphenophyte is similar to that of the lycophyte.

Pterophyta are the best known of the spore-bearing plants. They are the ferns. Most ferns are tropical, but many species live seasonally near streams and on moist forest floors. Some ferns are trees, but most are small, ground-dwelling organisms. The sporophyte generation is dominant in ferns, as it is in all vascular plants. The gametophyte generation is a thin, flat structure that is independent of the sporophyte. You may have seen the gametophyte structures on an indoor fern plant. They are often gray, hairy, flat structures that lie along the ground. The underground stem of the fern plant is called the *rhizome,* and the leaves of the fern are known as *fronds*. Starches and sugars are stored in the rhizome to keep the organism from dying during the winter. Come spring, the fern sends out fronds once again.

> **The sporophyte generation is dominant in ferns, as it is in all vascular plants.**

 **Exploration Activities**

1. What is the difference between a vascular plant and a nonvascular plant? How could you recognize a nonvascular plant if you saw one today?

2. Describe the reproductive cycle of a spore-bearing plant.

3. What do the spore-bearing plants have in common in terms of likely environments? Why?

## Spores and Seeds

One of the large steps in plant evolution was the appearance of seeds in gymnosperms and angiosperms. Seeds are small plant embryos, encased in the protective coatings of a fruit or a cone.

Every plant produces spores at some point in its reproductive cycle. The sporophyte generation begins with the zygote, which quickly grows into an embryo, then an adult plant that produces spores. In seed-producing plants, there are two types of spores—microspores, which grow into pollen grains, and megaspores, which grow into ova. In spore-producing plants, the spore becomes an archegonium, the male reproductive organ, or an antheridium, the female reproductive organ. In either case, the development of these organs begins the gametophyte, or sexually reproducing, generation. The gametophyte generation varies greatly in life cycle based on the plant.

At the end of the gametophyte generation, in either case, a fertilized potential plant is produced. Spore-bearing plants release spores, which carry chromosomes of the male and female plants, and these are borne to new places by wind, water, or animals. If a spore lands in a good spot, a new plant will grow.

> **Seeds are small plant embryos, encased in the protective coatings of a fruit or a cone.**

Most spores need to germinate quickly, or they will die. If drought conditions persist for a couple of days, a whole generation of spore-producing plants could fail to germinate. Spores often require moist ground or even water for reproduction.

Seeds, on the other hand, are released by the parent plant with a little insurance policy. They have a waterproof seed coat and a small amount of nutrients that can see them through bad times, even through harsh winters. In some gymnosperms, cones only open during optimal growth periods for the new, young plant.

> **Seed-bearing plants have developed a number of adaptations that transport seeds.**

On the whole, seeds offer a great advantage to a plant embryo. But there are disadvantages as well. Seeds are not transported as easily as spores. Plants spend a lot of available capital in providing for each seed, and so there are not as many seeds released as spores. Seed-bearing plants have developed a number of adaptations that transport seeds. You are probably familiar with the dandelion and its seeds. A puff of wind can move dandelion seeds several kilometers. Other seeds attach to passing animals or socks and hitch rides. Still others are covered by tasty fruit and depend on foraging animals to eat, move, and deposit them (with the added bonus of a little fertilizer!) into new locations. Some gymnosperms release their seeds only when a forest fire has prepared an optimal location for the young plants to grow.

 **Exploration Activity**

Create a cost/benefit analysis of seed and spore reproduction. If you were a plant, which reproductive strategy would you choose? Why?

> Gymnosperms are mostly evergreens, with long needles rather than flat leaves.

# Gymnosperms

The next group of plants include the phyla that are today considered to be part of the larger category of gymnosperms. Gymnosperm means "naked seed" and, unlike its later relations, produces seeds in cones or cone-like objects. Gymnosperms are mostly evergreens, with long needles rather than flat leaves. They are vascular plants and include the tallest and oldest living organisms on the planet.

In general, the gymnosperms form their seeds in cones. Like all plants, they have two generations. The spores are produced by the sporophyte generation. *Microspores* are produced in male cones, eventually giving rise to the gametophyte that develops pollen grains. *Megaspores* are produced in female cones, giving rise to female gametophytes, which produce ova. In gymnosperms, pollen is carried by the wind from the male cones to the female cones, producing embryos inside the female cones—seeds. The gymnosperms, unlike the spore-bearing plants, do not need water to reproduce. In plants, an embryo is the young, diploid sporophyte of a plant. Gymnosperms provide food-storage organs called *cotyledons* to support the embryo, and a tough seed coat that protects it until it germinates.

Some gymnosperms need an outside stimulus, such as a fire or summer heat, to open the cones. In fact, forest fire is a natural ally of large gymnosperms. Certain species, such as the giant sequoia and the coastal redwood, could not reproduce without it. Fire serves several purposes. It clears out the undergrowth near the trees, which allows young gymnosperm plants to get the sunlight they need, and the ash provides a natural fertilizer for the seeds and young plants.

Summer heat is another way to get the cones to open. You can try this yourself. Find a pine cone and place it near a warm, sunny window. Slowly, the cone will open. When you remove it from the sun's warmth, it will close up again just as slowly.

Coniferophyta, the conifers, are gymnosperms with seeds in cones. Pines, redwoods, spruces, and other cone-bearing trees all belong to this phylum. These trees often grow in mountains, where winter snows can be heavy. Trees in such environments risk breakage of

The ancient Red Dawn sequoia dates back to the Tertiary period of the Mesozoic era.

large limbs. Conifers, however, grow in a triangular shape, which allows snow to slide off harmlessly. The leaves, or needles, lose less water than broad leaves, and most conifers remain at least partially green year round, giving them a strong advantage in colder climates or higher terrain. (The only exceptions are the larches and bald Cyprus trees, which are *deciduous*.) There are over 600 species of conifers, including the largest living plants, the giant sequoia, and the oldest living things on Earth, the bristlecone pines.

Cycadophyta were common plants 250 million years ago. Today, because of changes to the environment, including temperature change and habitat destruction, there are only about 150 species left. Cycads are tropical plants and look very much like palms. However, palms are angiosperms; cycads are gymnosperms.

Gnetophyta are also a small group of species. Ephedra is one of the species and the only one that grows in North America. The other two species—*welwitschia* and *gneturns*, exist in Asia, South America, and Africa.

The smallest phylum of gymnosperm is the phylum Ginkgophyta, which has only one extant species, the *Ginkgo biloba*. This species is a native of China but has been transplanted elsewhere, and it has adapted to its new habitats. *Ginkgo biloba* is known for its large, fan-like leaves, which make it unique among the gymnosperms.

Ginkgophyta was once well-represented, but like all gymnosperms, their numbers and habitats are dwindling. Gymnosperms evolved along with the spore-bearing plants, but when the climate changed around 280 million years ago, many of the spore-bearing plants either died out or adapted. The gymnosperms thus became the dominant group of plants on Earth for a time. The continued drying of the land, along with the rise of the angiosperms, changed the balance yet again. Although they are by no means at extinction's door, the number of gymnosperm species has declined, while the number of angiosperm species has increased.

> **The number of gymnosperm species has declined, while the number of angiosperm species has increased.**

 **Exploration Activities**

1. What does the term *gymnosperm* mean?

2. You look out the window and see a pine tree. Does the pine tree represent the gametophyte generation or the sporophyte generation?

3. Why is forest fire a natural ally for large gymnosperms, such as the coastal redwood?

4. In addition to Coniferophyta, the conifers, what are the other phyla of gymnosperms?

5. Why are gymnosperms no longer the dominant group of plants on Earth?

# Student Lab: Tree Rings

Tree bark produces rings as trees grow. These rings enable us to determine a tree's age. A light layer is produced in the spring, when the growth of the tree is fastest, and a dark layer is produced in the summer when slower growth occurs. Tree rings also give clues about the amount of rainfall during given years. Narrow rings indicate slower growth and less rainfall. Wider rings indicate more rapid growth and greater rainfall. Wood is mostly made up of xylem cells, the tubular cells that conduct water from the roots to the leaves. Heartwood is found in the center of the tree and is made of xylem cells that no longer conduct water. Sapwood surrounds the heartwood and is made of xylem cells that still conduct water. Phloem cells conduct sugars made by the leaves to other parts of the tree. These cells are near the outside of the tree. Next to the phloem cells is cork.

## Materials

- Tree section or tree stump
- Crayons and paper

## Procedure

1. Make a rubbing of one-half of your tree section and label the bark, heartwood, sapwood, and rings.

2. Count the rings of your tree—a light and dark band make up one year. Check to see when wet and dry years occurred.

## Conclusion

According to the data you collected, how old is your tree?

# Angiosperms

All other plants on Earth are **angiosperms,** the flowering plants. What makes an angiosperm different from the rest? Flowers and fruits! Angiosperms (the phylum Anthophyta) are the largest of the plant phyla, with some 230,000 individual species. Like gymnosperms, angiosperms are seed-bearing plants. Also like gymnosperms, they have true root systems, vascular systems, stems, and leaves. Angiosperm seeds are formed when the ovule of the female part of the plant is fertilized by pollen from the male part of the plant. This reproduction occurs in the female reproductive organ, the ovary. Pollen comes from the stamen, which is the male reproductive organ. Both reproductive organs reside in the flower of the plant.

Once fertilized, most angiosperms produce some sort of fruiting body, which is the ripened ovary. Not all fruits are of the fleshy apple and watermelon variety. Most fruits are dry, and include things we call grains and nuts. The function of the fruit is to protect the seed. Fruits may also assist in the dispersal of the seeds. For instance, if a bear eats berries from a bush, it will obtain the nutrients and sugars from the berry. This encourages the bear to eat the berry. The seed, however, is also covered with a thick, waterproof coat, which is resistant to acids. More often than not, the seed passes through the bear's digestive tract and is excreted. A seed, or small group of seeds, with a ready-made fertilizer, either germinates immediately or winters over and germinates in the spring. Fruits are considered fleshy if they have a fruit wall full of sugars and water, and dry if they have a dry fruit wall. Different animals eat different types of fruit and have differing methods of dispersal. Other seeds disperse in other ways. Some fruits develop plumes that aid the seed in wind dispersal; others develop burrs that cling to fur and socks and move the seeds far from their parent plants.

Angiosperms took the world by storm when they emerged during the Jurassic period of the Mesozoic era, some 140 million years ago. Within a very short period of time, flowering plants nearly covered the earth. All grasses, hardwood trees, and virtually all plants thought of as "food" belong to Anthophyta, which is broadly categorized into two great classes.

The first class is the monocotyledons, or the **monocots.** Monocots have one seed leaf, which is a small leaf contained within the seed of the organism. Monocots are the smaller of the two classes and include families of grasses, bamboo, sugar cane, orchids, lilies, and palm trees. The second of the two great classes is the dicotyledon class, which has two seed leaves encapsulated within the seed. **Dicots,** as they are more generally known, represent all other types of plants—shrubs, wildflowers, strawberries, tomato plants, oak trees, grape vines, and everything else.

Angiosperms are extremely successful. They have special adaptations for living virtually everywhere on Earth. We will look at some of the specialized adaptations as we begin to look at the various parts of the plant. Some, such as root adaptations, also apply to gymnosperms, but others are specific to the angiosperms.

Angiosperms have one of three basic life cycles—*annual, biennial,* or *perennial.* Any gardener knows what this means. Annual plants live for only one growing season. They germinate, grow, flower, set fruit, and die in a single year. Most annuals are *herbaceous,* which means they do not have wood in their stems. By winter, the parent plant dies, but the seed set by the plant goes into a period of *dormancy,* a kind of suspended animation during which life processes slow to a near halt, and the seed germinates in the spring. Daisies are an example of an annual that you might see in your garden.

Biennial plants have a two-year life span. They spend the first year establishing a strong root system, but do not flower. In the winter, the aboveground portion dies back, but the root system has stored much, if not most, of the sugar the plant produced during the growing season. During the second year, the plant again grows, flowers, sets fruit, and dies. Carrots and turnips are typical biennial plants.

Perennials live for several years and sometimes many hundreds of years. They produce flowers and seeds once a year. Perennials survive the winter by entering a period of dormancy—they drop their leaves, for instance, or die back to ground level. The bulk of the woody stem, or the large bulbous roots underground, keep the plant alive during its dormancy. Trees and most shrubs are perennials, as are many species of grass. Your garden may have many herbaceous perennials, such as tulips, irises, and other bulb flowers, growing in it.

> **Angiosperms have one of three basic life cycles—annual, biennial, or perennial.**

> **Nearly all species grown commercially for food or to make fabrics are angiosperms.**

Angiosperms are extremely important as food sources for humans and other animals. Nearly all species grown commercially for food or to make fabrics are angiosperms. All grains, which form the staple diet of humans worldwide, as well as fruits and vegetables, are angiosperms. Only a few species of the other phyla—pine seeds, for instance, which are used in certain dishes, and the tender tips of some ferns—are used in the human diet. Angiosperms also play an important role in human medicine. Many modern medicines originated as herbal remedies from the dawn of humanity—most of the herbs are angiosperms.

Plants exhibit different kinds of tropisms. A tropism is a response to a stimulus of some kind. If the tropism is positive, the plant grows toward the stimulus. If the tropism is negative, the plant grows away from the stimulus. Plants grow toward light, which is known as *phototropism*. This is a positive tropism. Roots grow downward because of positive *gravitropism*. The stimulus, in this case, is gravity. Stems, on the other hand, exhibit a negative gravitropism, and grow away from the gravitational influence.

 **Exploration Activities**

1. What two bodies distinguish the angiosperm from all other plants?

2. When did angiosperms emerge in evolutionary terms?

3. Name the three life cycles of the angiosperm plant and explain what they are.

## Roots, Stems, and Leaves

Roots are the underground part of plants. Some roots are partially above ground, especially in very swampy areas, where the roots could drown and rot. Roots, even those in the ground, require aeration—they need to have air around them, or they will die, killing the plant. The root system anchors the plant into the ground and gives tall plants counterbalance, keeping them from falling over in strong wind. Roots absorb water and nutrients from the soil and transport these to the stem, where the xylem carries them upward. Many roots also store sugar and starch to keep the plant alive during periods of time when the plant cannot photosynthesize, such as in the winter months.

Within the root, water, nutrients, sugar, and starch travel via osmosis from cell to cell. The outermost layer of root cells is called the *epidermis*, which is the same name used for animal skin. The next layer inward is the *cortex,* used to transport water and nutrients from the outside of the root to the center. The inside of the root is called the *endodermis,* and it is from here that the nutrients begin their upward journey into the plant. Also within the endodermis are specialized cells that assist the root in making new root tissue that replaces damaged tissue. The layer that assists in regrowth is called the *pericycle*. The root also contains storage chambers called *parenchyma cells,* where water and sugars are stored.

When a seed germinates, the first visible object is a root, called the primary root. Smaller secondary roots grow from this root. Often, as the plant grows, the primary root is the main root for the plant. The mature root system consists of one, usually very long, *taproot,* or main root. Sometimes the taproot is longer than the stem! Secondary roots develop into branching *fibrous* roots. Unlike the taproot, which grows almost straight down, the fibrous roots do not grow deep into the soil. They branch out around the plant, giving it additional protection from wind and obtaining those nutrients that are likely to be closer to the surface, such as nitrates. At the very tips of roots, you will often see tiny roots called root hairs. These roots pull in the moisture and pass it along through osmosis to the epidermis. In fact, root hairs are made of epidermal tissue.

> **The stem is the transport system for the plant.**

The stem is a part of the plant that is above ground. Stems can be as large as the trunk of an oak tree, or as thin and frail as a morning glory stem. However large it is, the stem has several functions in any vascular plant.

First, the stem supports other aboveground organs, such as branches, leaves, cones, sporangia, and, in angiosperms, flowers and fruit. Some stems can carry out photosynthesis themselves, especially if they are herbaceous. Other plants have woody stems, such as trees and shrubs. We have already looked at the layers of a tree trunk, or stem, in a lab exercise. The stem is the transport system for the plant. Water and nutrients from the soil are taken up by the root system but are transported to leaves and other plant parts by the xylem, which is located in the stem of the plant. The phloem carries sugar, produced by the process of photosynthesis, back down the stem into the root system for storage.

The xylem is composed of *vessel cells,* which are tube-like cells with open ends. The phloem is made up of *sieve cells,* which are thin-walled cells with small holes in one end. A sieve cell does not possess a nucleus, but it has a companion cell that has one, and the companion cell regulates movement of sugars through the cell. Sugars move by osmosis from one cell to the next on their way down to the root system, to the fruiting body, or to repair the plant if necessary during the growing season.

The primary function of the leaves of any plant is to perform photosynthesis. Most angiosperm leaves are broad and have a lot of surface area where sunlight can be received. They are also flat, which allows sunlight to penetrate through the surface layer to the photosynthetic cells that lie beneath the surface layer of cells.

Leaves also regulate the amount of water contained within the plant through the process of transpiration. As mentioned earlier, the backs of leaves are covered with small pores called stomata, which are regulated by *guard cells.* When there is plenty of water in surrounding cells, the guard cells take in water and expand. As they expand, they bend inward, away from the stomata, and this opens the pore. When there is little water, the guard cells do not bend as much, and the stomata do not open as much.

> **Leaves have different shapes that are specially adapted for the region in which they flourish.**

Leaves have many different adaptations. Some leaves are specially designed for climbing, such as the modified leaves of pea plants and morning glories. Leaves have different shapes that are specially adapted for the region in which they flourish. Leaves with many "cutouts," such as an oak leaf, tend to live in areas where sunlight is freely available during the growing season. The cutouts help filter sunlight down to other leaves on the same plant. Leaves without such features tend to live in regions where sunlight is scarce, such as on the rainforest floor, where it is important to capture every ray.

In autumn, the leaves of deciduous trees, which lose their leaves, turn shades of red, gold, orange, and brown. In fact, these are the leaves' true colors. The process of leaf drop, which is called *abscission,* begins by the movement of nitrogen, minerals, and water from the leaves back to the root for winter storage. As the sunlight decreases in the fall, plants stop or slow photosynthesis. Chlorophyll fades from the leaf, and the leaf's true color—red, orange, yellow—can be seen. Leaves drop from the tree when an enzyme weakens the hold the leaf has on its branch by attacking the cell wall of the leaf stem.

 **Exploration Activity**

Pull up a root vegetable, such as a carrot or turnip, and submerge it in a container filled with water colored by a vegetable dye. Leave the plant for a few days, and watch the movement of the colored water through the root system into the stem and leaves of the plant. What do you notice?

# Flowers and Fruits

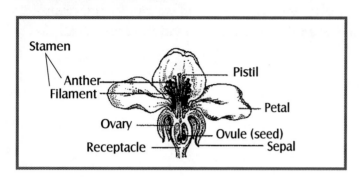

Sexual reproduction in angiosperms takes place in the flower, which is a complex structure made up of many separate parts. Some parts of a flower are the reproductive organs of the plant, while other parts entice pollinators to the flower. Although flowers look different from one another, they share a basic structure made up of four parts: **sepals, petals, stamens,** and **pistils.**

Petals are probably the easiest part to identify in most flowers. They seem a bit like leaves, but feel a little downier than a typical leaf. They are usually arranged in a circle, or series of circles, called a *corolla* at the top of a stem. Sepals are also like leaves. They are usually under the petals, and at one time, formed the bud of the flower. Inside the circle of petals, there are several protruding objects called stamens, which are the male reproductive organs of the flower. At the very tip of the stamen is the *anther,* which produces the pollen. Within the pollen are the sperm. At the very center of the flower are the pistils. These are the female organs. At the bottom of each pistil there is a slight enlargement of the organ; this is the *ovary* of the flower where the unfertilized ovules are formed and where the fertilized seeds will develop.

Some flowers have all four parts. They are called complete flowers. Some have only the male or female reproductive organs, and they are known as incomplete flowers. In any case, it is in the evolutionary best interest of the plant to *cross-fertilize*—for it to share DNA with others of its species.

Most flowers, although not all, have a mutual relationship with insects and birds that cross-fertilize. Some, like many grasses and grains, cross-fertilize by wind. Plants that do not need an animal pollinator usually do not have a visually appealing, delicately scented, nectar-filled flower. They often do not have a corolla and develop modified petals.

> **Each plant has a specialized dispersal system, including animal dispersal, wind, or clinging burrs.**

Nectar is a liquid power drink for insects, bats, and birds. It is made of sugars and proteins, and collects at the base of the petals. Animals are attracted by sweet odors from the flower, or by brightly colored petals. When the animal approaches the flower to drink the nectar, the animal brushes up against the anthers and collects some pollen, usually on its feet, feathers, or on "hairs" on an insect's body. When it leaves one flower and finds another, some of the pollen brushes off the animal's body and falls onto the pistils. When the pollen falls upon the pistils, the sperm have the potential of fertilizing the ovule within the ovary, and mitosis begins.

After fertilization takes place, the flower dies. Soon, the seed begins to develop. The wall of the ovule becomes a hard seed coat, which protects the young embryo. Inside, the zygote divides and undergoes mitosis; then, the ovary itself begins to develop into the fruit that will surround the seed. The ovary grows larger and, in the case of fleshy fruits, takes on a great deal of sugar and water from the plant. In the case of dry fruits, the ovary wall dries out. Each plant has a specialized dispersal system, including animal dispersal, wind, or clinging burrs. Plants that live near water produce seeds and fruits that can float.

The metabolic activity of the plant embryo drops dramatically during the period of time when the embryo is encased in the seed. Some plants are dormant only for a short time, while others need to survive a long winter. Seeds of plants that live in especially dry climates can remain dormant for many years.

When conditions are good, the seed germinates. The seed coat absorbs water just before germination. This typically happens in the spring, in environments where winter is an issue. Some seeds have a germination temperature requirement as well, needing temperatures in the soil as warm as 30°C before the seed begins to grow.

As the seed germinates, a tiny root called a *radicle* emerges. This will become the primary root and probably the taproot as well. As the root grows downward, the stem emerges and grows upward, toward the light. For an embryo, this is called the *hypocotyl*. The hypocotyl eventually breaks the surface of the soil, bringing along the initial seed leaf or leaves of the plant, which wither soon afterward. New leaves grow. The food stored in the seed leaves is soon used up as the plant begins making food on its own.

 **Exploration Activity**

Draw or create a visual representation of the life cycle of an angiosperm. Use whatever medium you wish.

## Kingdom Animalia

Kingdom Animalia is by far the largest kingdom of life on Earth, with 33 separate phyla and over 1.5 million defined species. Most of those species, some 95% of them, are *invertebrates,* animals without backbones. Many of those belong to the phylum Arthropoda, and of those, most are insects.

Animals have many different reproductive strategies and methods for obtaining food, and they live in specialized environments.

So what, exactly, do they have in common? Strangely enough, animals have only a few things in common with others of their kingdom.

Animals are multicellular and autotrophic. They must obtain nutrition from other living things in their ecosystems. At some point during their lives all animals undergo a period of mobility. Some animals are mobile throughout their entire lives. Others, such as mussels, attach to rocks and become *sessile.* But even in sessile species, the juvenile stage is *motile.*

The thing that defines the animal kingdom, more than any of these other traits, which they share with some other kingdoms, is that during embryonic development each animal develops a **blastula.** The blastula is a hollow ball of cells in a single layer that encloses a fluid-filled space. The blastula eventually forms the basis for the digestive cells, tissue, organ, or system of the animal, depending on the animal's level of organization.

Animal phyla are classified according to certain characteristics—level of organization, body plan, and symmetry. All complex animals have a true **coelom,** which is the body cavity of the animal. It is lined with *mesoderm,* tissue derived from the embryonic germ layer. The internal organs in a true coelom are more advanced than those in the simple animals. Phyla that have a true coelom and a digestive tract are divided into two large groups, the *protostomes* and the *deuterostomes.* As the blastula develops in the embryonic animal, an opening called the *blastopore* develops. In protostomes, the blastopore becomes the mouth, while in deuterostomes, the blastopore becomes the anus. Protostomes include the phyla

Mollusca, Annelida, and Arthropoda. Deuterostomes include the phyla Echinodermata and Chordata. Insects are in the phylum Arthropoda, sea stars are in the phylum Echinodermata, and human beings are in the phylum Chordata. Because of this division, even though we have more outward appearance in common with an insect than with a sea star, we are much more closely related to the sea star.

## Organization Levels

Animals are organized at the cellular level, the tissue level, the organ level, and the system level. Cellular level organisms are extremely simple. They do not possess a true coelom, nor any true organs. They have a sort of body cavity, but it is a mere space left over from the blastula stage of development. All activity takes place within the cells of the organism. An example of a cellular-level organism is the sea sponge.

Other organisms have collections of cells that form body tissues, such as muscle, but do not have any true organs. They are organized at the tissue level. Certain jellyfish fall into this category. They may possess certain cells that function as stinging cells, others that form tissues for expelling water, and others that form tissues for metabolism. However, they do not have cells that form organs. If you were, in the lab, to place a stinging cell and a metabolizing cell together, they would eventually form a single tissue of one or the other.

Other animals have organs, but the organs are separate entities that do not communicate with one another in any meaningful way. These animals are organized at the organ level. They have a true coelom, digestive organs, and perhaps organs capable of sensing prey nearby, but there is nothing linking the various organs. Sea anemones are organized at this level.

The most complex organization level is the system level. At the system level, animal organs are incorporated into larger systems—a stomach, small and large intestine, salivary glands, esophagus, and anus are all part of one large system, the digestive system, which communicates with the nervous system, circulatory system, and others. Many animals are organized at this level, including sea stars, insects, and human beings.

> **The most complex organization level is the system level.**

## Symmetry

Another way animals are often classified is based on their *symmetry*. Many animals have *bilateral* symmetry. The animal is symmetrical around a bisecting line down its center. This means that one side of the creature is a mirror image of the other. Humans, for instance, have two eyes, two nostrils, two ears, two arms, and two legs. Internally, there are other symmetrical organs—lungs, kidneys, and so forth. Phyla that have bilateral symmetry include Chordata, Arthropoda, Annelida, many of the phylum Mollusca, the Platyhelminthes, and many others.

Another form of symmetry is *radial* symmetry. In this form, the animal is symmetrical around a central disk. Echinoderms, who have radial arms projecting from the central disk, as well as Cnidaria, the phylum that contains the jellies, the sea anemones (pictured) and the corals, are radially symmetrical in their adult forms.

Some animals, usually very simple ones, are not symmetrical at all. Sea sponges, for instance, are asymmetrical.

## Body Plan

Some phyla have a characteristic body plan, or segmentation plan. For instance, all insects have a highly successful body plan that includes three major body parts—the head, thorax, and abdomen. Legs and wings, if any, are attached to the thorax. Antennae, eyes, and mandibles are located on the head. Most of the body organs are located within the abdomen. Consequently, it is very easy to look at an organism and determine whether or not that organism is an insect. Many other classes of organisms also have distinctive body plans.

 **Exploration Activities**

1. What is the single defining structure that makes an animal an animal? When does this structure appear?

2. What are the three ways animals are classified?

3. Name and describe the four levels of animal organization.

4. What is the difference between bilateral symmetry and radial symmetry?

5. Humans, like insects, have a distinct body plan. Describe it.

## Marine and Aquatic Invertebrates

The word *invertebrate* means "without backbone." Some 95% of all animal species fall into one of the 32 invertebrate phyla. We will look at several of the most populated phyla, in the water and on land.

The kingdom Animalia originated in the seas and is believed to have evolved from protoctists. Protoctist colonies folded inward, creating a *protoanimal*. In this early protoanimal stage, specialization of cells began to occur, making way for true multicellular organisms.

### Porifera

One of the first phyla of organisms that were truly animalian was the phylum Porifera, the sea sponges. Porifera are organized at the cellular level—they have no distinct tissues, organs, or systems—and they are asymmetrical. They are one of two phyla to be termed Parazoa—or, near animal. The other is the phylum Placozoa, which has only one species, a multicellular amoeba-like organism. Porifera have been on the planet a very long time. The earliest known fossils date from 550 million years ago. Porifera obtain nutrients by diffusion. Water enters the sponge through its many pores, bringing nutrients in and washing away waste products.

Sponges can reproduce both sexually and asexually. Sexual reproduction in the sponges involves a dispersal of sperm, which may or may not enter another sponge and fertilize an ovum. Most sponges reproduce asexually, when small pieces of the sponge break off and start growing on their own. In early life, sponges are ciliated larvae, which are free-swimming. As they reach the adult stage, they attach themselves to rocks and become sessile.

The lavender tube sponge

### Cnidaria (Coelenterates)

Almost all Cnidaria are marine animals. This phylum includes jellies, sea anemones, and corals. They are radially symmetrical.

> Some 95% of all animal species fall into one of the 32 invertebrate phyla.

Most Cnidaria are organized at the organ level. They have a single body cavity that has one opening, which serves as both mouth and anus for the organism. They have a nerve net, capable of stinging prey it touches, but no nervous system or blood vessels. All Cnidaria are carnivores, and they all sting prey into submission, then carry it to the body cavity opening.

Some Cnidaria, such as most hydras, most corals, and most colonial hydroids, reproduce asexually. They bud off, and the new organism is an exact duplicate of the parent. However, all Cnidaria can also reproduce sexually if necessary. The sperm and ova are released into the water, where they fuse and form blastulae, which develop into free-swimming larvae called *medusas*. Some Cnidaria never become sessile; others spend the majority of their lives in the sessile condition. A few species of jellyfish can be extremely harmful, even fatal, to human swimmers.

## Mollusca

Mollusks are soft-bodied animals, usually possessing an external or internal shell. About 110,000 species have been described, making Mollusca the second largest phylum in the kingdom after the arthropods. Mollusks live in both marine and freshwater environments. Mollusks are bilaterally symmetrical. They are fairly highly organized and have digestive and circulatory systems. Mollusks grow their own shells using a special organ, called the *mantle,* to construct their hard shells out of calcium carbonate in the water. Mollusks also have a "foot" that emerges from the shell to help them move. There are eight classes of Mollusca. The three largest classes are Cephalopoda, which includes tentacled mollusks, such as the octopus and squid; Peclecypoda, also known as the bivalves, which includes mussels, oysters, clams, and scallops; and Gastropoda, which are single-shelled mollusks, such as snails and whelks. Most bivalves eat by filter-feeding. They live in sandy bottoms or attached to tidepools and wait for the ocean to bring their meal to them. Gastropods and cephalopods sometimes hunt, actually trapping small crustaceans, fish, and other zooplankton.

The sexes are separate in mollusks, although some mollusks can reverse their sex several times within the mating season. Fertilization, except in cephalopods, takes place in the water. Male cephalopods use a tentacle to place a sperm packet inside the female's

> **Mollusks are soft-bodied animals, usually possessing an external or internal shell.**

mantle cavity. Female cephalopods look after their eggs until they hatch. Once hatched, mollusk young are unprotected until they form their first shell. They float with the rest of the zooplankton in the ocean. Once they grow a shell, they settle to the bottom of the shallow sea or attach to a rock for the rest of their lives.

## Echinodermata

Echinodermata, animals with spiny skin, are highly organized, with digestive, circulatory, and simple nervous systems. Although radially symmetrical animals at the adult level, echinoderm larvae are bilaterally symmetrical. All species are marine. They include sea stars, sand dollars, and sea urchins, and they make their homes in tidepools and shallow ocean areas. The echinoderms are omnivores, scavengers, and predators. They eat by filtering sea water, grazing on the algae beds, eating dead and decaying material, and direct hunting. They will eat everything from small zooplankton and phytoplankton to large clams. Sea stars, especially, are able to use specialized tube feet to pry open bivalves, inject their own stomachs into the bivalve, eat the meat, and pull their stomach back into their own bodies. Sea stars are the major predator of the intertidal zone. Echinoderms have a water vascular system that enables them to move, capture prey, and cling to rocks during tidal changes. By decreasing the amount of water in the organism, the organism's tube feet can create an incredibly strong hold on a clam shell or tidepool wall.

> Echinoderms have a water vascular system that enables them to move, capture prey, and cling to rocks during tidal changes.

Echinoderms reproduce by spawning; they shed sperm and ova into the water, and fertilization takes place there. When the eggs hatch, the larvae join the ranks of the zooplankton until they *metamorphose* into adults. Most echinoderms can regenerate lost limbs; some can even regenerate disks. Some echinoderms, such as the sea cucumber and sea urchin, are used in the human diet as sushi.

## Platyhelminthes

This phylum includes the flatworms, which are organized at the organ and

Sea star, commonly called starfish

> **There are about 15,000 species of flatworms.**

system level, and are bilaterally symmetrical. Some independent species are marine, but the vast majority of free-living species are freshwater aquatic. There are about 15,000 species of flatworms. Some species are parasitic, including the liver fluke, which parasitizes an animal host's liver, and the tapeworm, which lives in the intestinal tracts of animals. Other species are free-living. The parasitic species are adapted to live on specific animal organs or tissues; the free-living species are carnivores and scavengers. They eat fairly large prey, relative to their size, such as insects and crustaceans. The flatworm has an organ-level digestive system and a very simple nervous system, which is composed of two eyespots at the top of its head, and nerve cells that extend down the length of the creature from each of the eyespots.

Flatworms are remarkable for their ability to regenerate. A classic biology lab class involves slicing a flatworm down the center, along the bilateral line, and watching for several days while each half regenerates. This is asexual reproduction in the flatworm. Most free-living species have both male and female sexual organs—they are *hermaphrodites*. This means they are able to cross-fertilize with another organism, a process in which each organism fertilizes the other. The exchange of genetic material occurs internally, and the flatworm lays fertilized eggs in a cocoon-like object. Tapeworms, alone among the flatworms, tend to be self-fertilizing. Most free-living flatworms produce eggs that hatch into small adults, while parasites tend to have life cycles that include a larval stage.

 **Exploration Activity**

Solve the crossword puzzle below with words from this section.

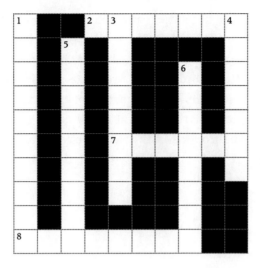

**Across**

2. a tentacled mollusk

7. a type of symmetry around a central disk

8. phylum containing shellfish, such as clams and whelks

**Down**

1. a sea star is an example of this

3. phylum containing jellyfish and sea anemones

4. a lifestyle in which an animal stays in one place

5. a type of symmetry around a bisecting line

6. phylum containing sea sponges

# Terrestrial Invertebrates

Just as in the water, the majority of all animals on land are also invertebrates. The largest phyla that live primarily on land include Arthropoda, the insects, arachnids, and crustaceans; Nematoda, the roundworms; Annelida, the segmented worms; and some members of Mollusca, especially terrestrial slugs and snails. We have discussed mollusks at some length already, and we will cover arthropods in the next section. Here, we will primarily discuss roundworms and segmented worms, which include familiar species of earthworms.

## Nematoda

Roundworms are widely distributed and can live in soil, as well as in both marine and freshwater environments. Roundworms include many free-living species and also many parasitic species, including some parasites that attack humans and domestic animals. Roundworms are bilaterally symmetrical and are organized at a very simple system level. They are the simplest organism to have a full digestive system. Free-living species also have very highly developed sensory organs. It has been estimated that a third of the human population of the world has some health problem associated with roundworms. Examples of nematodes that can be harmful to humans include hookworms; pinworms, the worm that causes trichinosis, which can be ingested in humans by eating the muscle meats of undercooked pork; and Rhabditis, which infests the root systems of garden plants, causing them to eventually lose their fruit. Another species causes heartworm in dogs. However, free-living nematodes are essential to a healthy ecosystem, since many of them are primarily decomposers.

Roundworms mainly reproduce sexually and lay eggs, which hatch into small but mature organisms. Roundworms are hermaphrodites and have both sets of sexual organs in their bodies.

> **Roundworms are widely distributed and can live in soil, as well as in both marine and freshwater environments.**

## Annelida

All members of this phylum are segmented worms. Members include earthworms, bristleworms, and leeches. Their name means "small rings," and when you observe an annelid, you see ring-like segments. The segments are not only skin-deep. They continue all the way to the center of the animal. Each segment has its own muscles, which allow the worms to move in their characteristic fashion as the segments perform a coordinated series of muscular contractions and releases. The segments also allow for specialization of functions. Most contain excretory organs, which aid in motion, and nerve centers. But only a few contain digestive organs or reproductive organs. Annelids are bilaterally symmetrical and are organized at the system level, possessing complex digestive, circulatory, and nervous systems. Annelids are coated in a waxy cuticle, which offers them some protection against predators and cold ground.

Members of the phyla live in moist, unfrozen soil nearly everywhere on the planet or, in the case of bristleworms and leeches, in unfrozen ponds, lakes, and even oceans. They are not found in soil in the desert, nor are they found in the seas under the polar ice caps. Segmented worms are important decomposers in their ecosystems and are also important in aerating the soil, allowing oxygen to reach plant roots. Annelids are hermaphrodites. Earthworms and their near relatives fertilize ova internally; the marine and aquatic species release ova and sperm into the water.

> **Segmented worms are important decomposers in their ecosystems and are also important in aerating the soil, allowing oxygen to reach plant roots.**

 **Exploration Activities**

1. What kinds of problems can humans experience from parasitic roundworms?

2. How far into the organism does segmentation go?

3. Is a leech an annelid or a nematode?

## BACKGROUND

# Arthropods

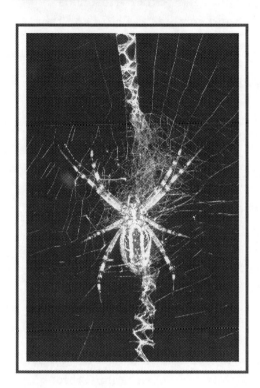

By far, the largest phylum of life on Earth—of any kingdom—is the phylum Arthropoda. At least 1.5 million arthropod species have been described so far—and biologists believe that once all living things in the rainforests are accounted for, arthropods will top 10 million species. Arthropods include insects; arachnids, such as spiders and scorpions; crustaceans, such as lobsters, crabs, shrimps, and crayfish; millipedes and centipedes; and horseshoe crabs.

## In Common . . .

The word arthropod means "jointed foot." All arthropods have jointed appendages. These are useful because they allow for strong and diverse motion during the movement of the animal, and can be used in ways other than locomotion. For instance, spiders use their jointed legs to capture prey, repair webs, mate, and sense surroundings. Some insects use their legs to make a characteristic sound, such as a cricket chirp, by which they call to potential mates. The large claws of crustaceans, actually modified legs, are used for defense.

Arthropods also possess a hard **exoskeleton**, which is a thick coating of chitin, the same substance that makes up fungal cell walls and human fingernails, as well as some proteins. In many species the exoskeleton is a continuous body covering. In crustaceans the exoskeleton is made up of plates. Exoskeletons protect internal tissues and organs, and provide places for the muscles of some species to attach.

Exoskeletons provide excellent protection, but they have numerous disadvantages. They cannot grow, and must be *molted*—shed—from time to time as the organism matures. It takes a while for the exoskeleton to grow back, and during that time, the animal is defenseless. Exoskeletons are made of heavy material, and since many arthropods fly, they have a lighter exoskeleton. This still protects body organs, but it does not offer the same level of protection from predators.

> By far, the largest phylum of life on Earth—of any kingdom—is the phylum Arthropoda.

All arthropods also show segmentation, culminating with the remarkable body plan of the insect class. Insects developed a fused segmentation system, in which they are left with three main body parts—a head, a thorax, and an abdomen. Other arthropods continue the fusion still further and combine the head and thorax into one body segment.

## Organization, Symmetry, and Systems

Arthropods are all bilaterally symmetrical, system-level organized animals. They have evolved three separate respiratory systems, which make use of three different types of respiratory organs—*gills, book lungs,* and *tracheal tubes and spiracles.*

Gills are found in many aquatic species, including the crustaceans. Aquatic arthropods exchange gases with the water itself. They take in dissolved oxygen and expel carbon dioxide through a series of gills, via osmosis. When water passes over the gills, gases are exchanged with gases in the blood.

Spiders and their near relatives have book lungs, which are folded membranes that look much like a closed book. Air enters the chamber where the book lungs are from the outside of the animal's body and is diffused into the bloodstream of the animal through these membranes.

Insects have tracheal tubes, which are a branching network of air passages that carry the air throughout the insect's body. Air enters and leaves the tracheal tubes through openings in the insect's body called spiracles. A finely tuned muscle system helps to pump the air in and out.

Arthropods also have a well-developed nervous system. The system consists of a nerve cord attached to a brain, and several *ganglia*—bulb-like enlargements of nerve tissues. Arthropods' nervous systems are attached to the many sensory organs. Most arthropods have eyes on their heads. The eyes are either simple, in the case of spiders and their relations, or are *compound,* which allows the animal to see in many directions at once, perceive color, and detect the slightest movement. They also may have antennae, or modified antennae known as *pedipalps,* which are stalk-like structures that detect changes in the environment and are used by many species for intra-species communication. Antennae can also

> **Arthropods are all bilaterally symmetrical, system-level organized animals.**

sense the odor of *pheromones,* which are chemical odor signals given off by all animals. These are important, biologists believe, in the sensing of trails, such as those followed by ants and butterflies, as well as in mating.

Arthropods have a complete circulatory system, with a heart and blood vessels, as well as a complete digestive system, including a mouth, stomach, intestine, and anus. Some arthropods excrete via an organ called a *Malpighian tubule.* Arthropod *mandibles,* or jaw-like structures, are specifically adapted to the lifestyle of the particular organism. They can hold prey, sponge up fluid, suck nectar from flowers, inject venom, or draw blood from other animals. Arthropods also have a well-developed muscular system. Muscles are attached to the exoskeleton on both sides of each joint.

Arthropods reproduce sexually, by usually separate male and female organisms (a few are hermaphroditic). Mating of terrestrial organisms usually involves a mating ritual or "dance." Land arthropods usually fertilize internally, and the female lays eggs. Marine or aquatic arthropods reproduce by spawning ova and sperm into the water, where fertilization takes place. A few species can reproduce asexually via *parthenogenesis,* in which a new organism develops from an unfertilized ovum. The new organism is invariably female and sterile. Ants and bees are well known for this. The queen lays many eggs, but the only fertile ones are the ones that result from sexual reproduction. The others become workers, females that are almost an exact genetic duplicate of the queen.

> **Arthropods have a complete circulatory system, with a heart and blood vessels, as well as a complete digestive system.**

 **Exploration Activities**

1. What structure gives arthropods their name? How are these structures used?

2. What are the benefits and drawbacks of an exoskeleton?

3. Name the three respiratory systems possessed by arthropods, and speculate as to the reason the different classes have differing systems.

4. Arthropods have several well-defined systems. How are their systems different from those of higher organisms, such as humans or birds? How are they the same?

**BACKGROUND**

# Arthropod Evolution and Diversity

Arthropods are found in many ecosystems around the world, because their remarkable body plan is adaptable to many habitats. Arthropods exist in commensal relationships, mutual relationships, and parasitic relationships wherever they are found.

Most likely, arthropods descended from annelids. As they evolved, the body segments of the annelids fused into a fewer number of body parts and developed more organized systems. The exoskeleton of the arthropod is much hardier and offers more protection than the waxy cuticle that surrounds the annelid body.

Because the hard body parts of arthropods often fossilize in some way, a great deal is known about their evolution. An ancient arthropod class, the *trilobites,* ranged in size from extremely small to several meters in length. All trilobites died out suddenly at the end of the Paleozoic era. Large arthropods that have modern descendants populated the warm, swampy lands of the Carboniferous period. Some dragonflies had wingspans of two meters, and millipedes were often 3–4 meters in length. Scorpions were at least 1 meter long.

As the great fern forests died out, and the climate changed to a drier, cooler environment, the arthropods also changed to roughly the size they are today. Some small relics of the Mesozoic era were caught in amber and are perfectly preserved even today.

There are six existing classes of arthropods. The most ancient members are the horseshoe crabs, the Merostomates, which have only four extant species. Horseshoe crabs look like ancient masks with a long spike at one end. They are marine and emerge from the ocean only to reproduce. These ancient creatures date from the Paleozoic era and are virtually unchanged from their ancestors.

Diplopods and Chilopods are the millipedes and centipedes, respectively. Centipedes are carnivorous and will eat smaller arthropods, mollusks, worms, and any small animal they can catch. They occasionally bite humans, and their bite is painful, but not poisonous. Millipedes are mostly herbivorous and fairly harmless. They emit a nasty-smelling fluid from specialized glands if

> Because the hard body parts of arthropods often fossilize in some way, a great deal is known about their evolution.

disturbed and have a nasty flavor that encourages most other animals to leave them alone. Both classes have tracheal tubes for respiration, like insects.

Arachnids include spiders, scorpions, ticks, and mites. All arachnids are carnivorous and contain venom, but most are harmless to humans. Only 12 of the more than 30,000 species of spider are poisonous enough to make a human sick, and even those few do not usually kill an animal as large as a human being. In North America, only the black widow spider and the brown recluse are so venomous that they can cause noticeable illness. Almost all scorpions have a stinger in their tail that they use to immobilize prey. Some species of scorpion venom is very dangerous to human beings, but most of those species live in Africa.

Arachnids have two distinct body segments—a fused head and thorax called a *cephalothorax,* and a large abdomen. Many spiders have silk glands and spinnerets for weaving webs; others dig holes in the ground and wait for prey to happen by. Some arachnids are large enough to immobilize and consume small vertebrates such as birds and mice. Their mouth parts have formed fang-like structures called *chelicerae.* Arachnids have eight pairs of legs.

Arachnids are found virtually everywhere on Earth, from the deepest rainforest, to the tops of the highest mountains, to the driest deserts on the planet. Few live in water. Some, such as ticks, are more dangerous than outwardly more poisonous arachnids. Ticks are a vector for many human diseases, such as Lyme disease, Rocky Mountain spotted fever, and others.

Most crustaceans are mostly or entirely aquatic, with the single exception of the pill bug. Even this small land crustacean must live near water, because, like other crustaceans, it exchanges gases through gills, rather than book lungs or tracheal tubes. Other members of the class include crabs, lobsters, crayfish, shrimps (including the tiny krill that form the bulk of the diet of the world's largest animal, the blue whale) and barnacles. Many people believe barnacles are members of the Mollusca phylum. Indeed, they look alike, with hard shells that resemble clams or mussels. Barnacles are very different from most crustaceans. Unlike the others, barnacles are sessile. They anchor themselves to rocks, ships, and other firm surfaces, and filter seawater to obtain nutrition.

All other crustaceans are bottom feeders. They are important in the marine and freshwater ecosystem, because they clear the sea floor of dead and decaying matter. Occasionally, during particularly bad times, crustaceans will catch and eat small animals. They catch the prey with one of their pinchers and pull it apart with their mandibles. Most crustaceans have five pairs of legs, and their two compound eyes are usually on stalks.

Insects are the largest class in the phylum. They live virtually everywhere on the dry surface of the planet, and there are several aquatic species as well. Insects often undergo a series of changes in body structure. This is not the case with the other members of the phylum, who are hatched as miniature adults.

Most insects hatch into a *larval,* or juvenile, stage. This stage is free-living, wingless, and often worm-like. The organism is sometimes called a caterpillar at this stage, whether it will grow up to become a butterfly or not. There are specialized names for some insect larvae—most fly larvae are known as maggots, for instance. As the young creature eats and grows, it molts several times.

Next, the insect enters the *pupa,* or changing, stage. During this stage, the insect has built a cocoon or other structure around itself. Some tissues are broken down, others are built up. After a time, the adult animal emerges, mates, and begins the cycle again. This cycle is called *metamorphosis.* When an insect goes through all the stages, it is known as complete metamorphosis. Butterflies, for instance, engage in complete metamorphosis.

It is an advantage to the species for metamorphosis to occur, since the adults and the juveniles are not in competition for the same food sources. Caterpillars feed on the leaves of the tree they were born on, for instance, while butterflies drink the nectar of flowers.

Not all insects go through complete metamorphosis. Some merely molt a few times until they reach adult size. The young of species that undergo incomplete metamorphosis are called *nymphs.* Nymphs are typically wingless and do not reproduce, but they look very similar to adults, and they are in competition for resources.

> **Insects often undergo a series of changes in body structure.**

## Exploration Activity

Write an essay about how the various classes of the phylum Arthropoda resemble one another, and what their major differences are.

# Sea Squirts and Invertebrate Chordates

The chordates most familiar to us are the vertebrate chordates, the animals with backbones. However, our earliest chordate ancestors looked more like the animal in this photo than any vertebrate, living or extinct, that you can think of.

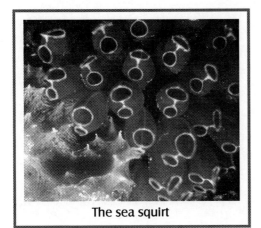
The sea squirt

These animals are sea squirts. They take their name from the way they squirt out sea water when taken out of the ocean. They belong to the phylum Chordata, or animals in possession of a **notochord.** During the embryo stage of all vertebrates, this cord becomes the backbone. In invertebrates, the notochord stands alone.

When sea squirts are larvae, they are motile. They use the notochord to propel through the water. The larval sea squirt resembles a young tadpole and has gills to obtain oxygen. It also has a complete digestion system and a nervous system that includes the notochord and a *dorsal nerve cord*. The larval phase ends quickly for the sea squirt, or as it is properly known, the *tunicate*. After just one day, the larva attaches to a rock or other hard surface and goes through a period of metamorphosis, during which the notochord and nerve cord are absorbed, and two openings, called *siphons,* grow. The siphons are coated with cilia, which cause a minor current to enter the openings. This allows the sea squirt to filter feed and to obtain oxygen from the water through the gills, which are now located within the siphon.

Recently, a DNA survey of the sea squirt revealed that the animal was surprisingly similar to the higher vertebrates. The suspicion that sea squirts and higher vertebrates share a not-too-distant common ancestor is not new. Charles Darwin suggested that the sea squirts and vertebrates diverged from that common ancestor around 550 million years ago.

Another invertebrate chordate is the lancelet, which is a fish-like organism that retains its notochord throughout its life. It lives under the sand most of the time, with only its mouth protruding, waiting for unwary zooplankton to happen along.

> **Charles Darwin suggested that the sea squirts and vertebrates diverged from a common ancestor around 550 million years ago.**

 **Exploration Activities**

1. What phylum do sea squirts and lancelets belong to?

2. How does the respiratory system of the sea squirt change as the animal matures?

## BACKGROUND

> A mere 5% of all living animals are vertebrates.

> From amphibians, both mammals and reptiles evolved on separate, parallel evolutionary paths.

# The Vertebrate Classes

In addition to the normal order of classification, from time to time more divisions are necessary. These are sometimes called sub or super classifications. One important example is the subphylum Vertebrata, which is part of the phylum Chordata, and includes all chordates with a hard backbone. There are five groups of vertebrates. They are fish, amphibians, reptiles, birds, and mammals.

A mere 5% of all living animals are vertebrates. Vertebrates have certain advantages, offered by an internal skeleton, that the invertebrate phyla do not possess. Vertebrates do not need a heavy exoskeleton to protect their body organs. Most vital organs are contained within a cage of bones in the main body trunk of the animal. An internal skeleton allows for growth, as well. Bones, like any other internal tissue, can grow without having to be replaced. Vertebrae are also more flexible than exoskeletons, making locomotion much easier for the vertebrate classes.

Like anything in nature, there are drawbacks as well. Internal bones do not offer a good protection against predators, so vertebrate species developed other means of defense, including banding together in schools or herds, evolving protective coloration or defensive structures such as horns, or, in one special case, learning to control fire and make tools. Bones can break, leaving the organism helpless. In the wild, a broken bone means certain death. A single fall from a tree could kill a squirrel; the same fall for a much smaller beetle would be a minor inconvenience.

Vertebrate classes are all bilateral and system-level organized. They all reproduce exclusively by sexual reproduction. They are primarily classified based on a few important factors—the number of chambers in their hearts, whether or not they produce amniotic eggs, and whether they are endothermic or exothermic.

Vertebrates evolved as fish some 400 million years ago, during the Devonian period of the Paleozoic era. Amphibians eventually evolved from fish that had developed lungs, and were able to leave the sea for periods of time. From amphibians, both mammals and reptiles evolved on separate, parallel evolutionary paths. Birds are thought to have evolved from reptiles during the Mesozoic era.

 **Exploration Activity**

Do some research, then create a *cladogram,* or branching family tree, for the subphylum Vertebrata. Note that some vertebrate superclasses, such as the superclass that contains all fish, have more than one actual class. Fish, for instance, are classified into three classes, even though they are part of the same superclass.

**BACKGROUND**

> Fish are entirely aquatic animals that live in either oceans or fresh water.

# Fish

There are more fish species than all other vertebrates put together—more than 30,000 species. Fish are entirely aquatic animals that live in either oceans or fresh water. Their respiratory system is based on gills, although a very few transitional species also have lungs and can survive periods of time breathing oxygen from the air. While breathing with gills, which are filaments that contain tiny blood vessels, fish take in water by mouth. This water passes over their gills and exits through the gill slits on the sides of the fish. Oxygen and carbon dioxide are exchanged in the small blood vessels through osmosis.

Fish have a two-chambered heart. One chamber receives blood from which the cells have removed all the oxygen, while the second chamber pumps the blood directly to the capillaries of the gills, where the oxygen is collected. Oxygen is then carried in the blood to the tissues of the animal. Blood flow in a fish is sluggish because the capillaries of the gills are very small, and it takes time to move all the blood through those tiny veins.

Fish live in virtually every aquatic body, except in those that are too salty to support eukaryotic life. They live at every depth, as well, and have developed interesting and complex adaptations for living at various depths in the ocean. Fish can be very tiny or very large. Some sharks are longer than 15 meters.

The fish are separated into three great classes—the agnathans, or jawless fishes, which include lampreys and hagfishes; the chondrichthyes, or the *cartilaginous* fishes, whose members include sharks and rays; and the osteichthyes, or bony fishes, which include sea horses, eels, tuna, goldfish, and all other fishes.

The agnathans have no hinged jaw. They have long, thin, tube-like bodies, with no paired fins and no scales. Hagfish survive on dead or dying fish, by attaching to them and sucking out blood and organs through a slit mouth ringed with many sharp teeth. Lampreys are parasites and attach themselves to other large organisms by a sucker-like mouth. Both families have skeletons made of **cartilage.**

The chondrichthyes, or the cartilaginous fishes, possess ordinary-looking skeletons composed of cartilage. They have paired fins. Fins are fan-like membranes that are supported by stiff spines called rays. Fins have different functions, depending on where they are located on the fish. Some fins keep the animal upright in the water, some provide locomotor capability, some move the fish in the direction it wishes to go, and some are instrumental in reproduction.

Not all fishes have all the different kinds of fins. Dorsal fins, located on the back of the fish, and anal fins, located on the *ventral* (underside) side of the fish and toward the tail, keep the animal from rolling from side to side while swimming. The *caudal* fin, located on the tail, is used for locomotion, propelling the fish forward in the water. The *pectoral* and *pelvic* fins, located on either side of the fish, assist the animal in steering. The pelvic fins on male fish also hold the female steady for internal fertilization. In the chondrichthyes, the first jaws developed. An example is the jaws of a shark, which can open and close around the shark's prey.

Sharks and their near relations have developed an excellent sense of smell, by which they can track injured animals many kilometers under water. Another adaptation that helps higher fish to sense their environments is the *lateral line,* which is a visible line of fluid-filled pits that detects movement and vibrations in the water.

Chondrichthyes also have scales, which are overlapping bits of cartilage that cover the skin of the animal, offering protection from predators and environmental conditions. Chondrichthyes have external gill slits—they are visible on the outside of the animal. This class, unlike many of their relatives, have internal fertilization. All of these fish are capable of laying eggs, but sharks may produce as few as 20 eggs and keep them within their bodies until they have hatched and grown to a substantial size (about 40 cm). These young, when "born," function like small adults, and many, if not most, survive.

Sharks are an example of an animal with a reproductive strategy that is more like our own, producing fewer offspring, but seeing the majority of them survive. This is known as a K reproductive strategy. The opposite is the p reproductive strategy, in which animals produce millions of eggs and trust that of those millions, a few might survive to adulthood. Unlike animals such as the

> **Sharks and their near relations have developed an excellent sense of smell, by which they can track injured animals many kilometers under water.**

mountain gorilla, which might have one offspring every six or seven years, sharks are not at the extreme K level. Each organism falls somewhere on the line between extreme K and extreme p.

The final class is osteichthyes, the bony fishes. Each of these organisms has a skeleton made of bone. Bone is a mineralized living tissue, usually of some sort of calcium, that makes up the endoskeletons of most other vertebrates. Bone was an important evolutionary advance, since the more flexible and sturdier bony structure allowed fish to adapt to many different aquatic environments, including a transition to land. Separate vertebrae, such as we see in bony fish and other vertebrates, allows for greater speed underwater.

Like the other fish classes, osteichthyes also have gills, highly developed sensory organs, and paired fins.

Unlike the other fish classes, osteichthyes evolved a new organ called the swim bladder, which helps fish control their depth. This allows fish to live in the deepest oceans, among coral reefs, or in tiny tidepools. Fish with a swim bladder can adjust the amount of air contained within the structure. Less air in the swim bladder means the fish is less buoyant and can swim at deeper depths. More air means the fish stays near the light zone of the ocean, perhaps to feed on phytoplankton.

In the bony fishes, the sexes are separate, and fertilization usually occurs by *spawning*. During spawning, some female fish lay many millions of eggs. Such fish, in contrast to the shark, find themselves at the p end of the reproductive-strategy continuum. A few bony fish give live birth. In these few species, many of which are aquarium pets—guppies, mollies, swordtails—fertilization is internal. The young fish develop within their mother's body and are born when mature.

Some bony fish are transitional species. Lungfish, for instance, possess both gills and lungs, and can live for short periods of time out of the water. It is currently believed that lungfish, or their relatives, the lobefish, were the direct ancestors of amphibians.

 **Exploration Activities**

1. Compare how jawless fish and cartilaginous fish feed. Why was the development of jaws such an important step in fish evolution?

2. What is the lateral line, and how does it assist fish in survival?

3. Why do fish require so many different types of fins?

4. Why was the evolution of a swim bladder so important?

# Student Lab: How Fish Swim

As discussed in this section, fish use fins to move through the water. There are several ways fish use fins for locomotion. Most fish move in one of three ways. Some swim in an S-shaped pattern, which requires contracting and releasing muscles on either side of its body in a coordinated manner. A clear example of this sort of locomotion is in the eel, whose paired fins are very small. Another swimming style is to keep the front portion rigid while moving the tail and back of the body in strong strokes. Mackerel and other long, thin fishes use this technique. The third style is to keep the entire body rigid and move only the tail. Tuna and other large fishes do this. This method moves the fish through the water faster than any other method.

## Materials

- Fish tank, trip to local aquarium, or nature video featuring fish
- Field guide to identify fish species

## Procedure

1. Carefully observe several different species of fish, from the top of an aquarium if possible. Determine which swimming patterns they follow.

2. Identify each species by consulting your field guide. If you are present during feeding, note any changes in swimming style during feeding.

## Conclusion

Compare the swimming patterns of the fishes you observed. What might be the advantage to the fish of this style of swimming in terms of food or escape from predators?

**BACKGROUND**

> Their name means "double life," since amphibians spend the first part of their lives in the water and must return to the water to reproduce.

# Amphibians

Vertebrates took a fateful step—literally—about 350 million years ago. They began, slowly and tentatively, to move from the water of the world onto dry land. Conditions in the seas had changed dramatically since the first animal life originated there. Because of the incredible profusion of life, there was serious competition for resources, and not just food resources. Oxygen, too, was being depleted in the oceans and in millions of small ponds and lakes around the world—it was being used up by the animals faster than it could be replaced by bacteria, algae, and new water plants. With thousands of carnivorous species, it was also a dangerous place. Any fish that could get out of the water for a little while had a better chance of survival. Plus, periods of drought were drying up small ponds worldwide.

The first *amphibians*—animals that lived both in water and on land—were most likely the direct descendents of the oxygen-breathing fish that left drying ponds to look for new water sources. Their name means "double life," since amphibians spend the first part of their lives in the water and must return to the water to reproduce. For the first amphibians, life on land was a good prospect. There were plenty of arthropods and other small organisms on which to feed, and the oxygen was more plentiful. Plant life was firmly established on land and was emitting vast quantities of oxygen into the air. The earliest amphibians developed legs from the fins of the lungfish that were their ancestors, which made locomotion somewhat easier.

Life was tricky on land for a couple of reasons. Amphibians, like fish, are *ectothermic*. While the temperature of water stays fairly constant, this is not true of air temperature, which can fluctuate greatly. Amphibians also had to return to the water to reproduce, and in some cases, to breathe.

Amphibians, in the adult form, have lungs, but they absorb oxygen through their skin, which must remain moist in order for osmosis to occur. Amphibians do not possess scales, claws, or even eyelids, which makes long-term land life difficult. Also, amphibians, like fish, lay eggs that have no hard shell, and the larvae obtain nutrients and oxygen that are dissolved in the water through diffusion.

> Amphibians go through a period of metamorphosis.

Amphibians go through a period of metamorphosis. They emerge from their gelatinous egg as tadpoles, with gills and no legs. Over a period of time, amphibians slowly lose the long tadpole tail and begin to grow legs. The lungs develop, and finally, the young frog or salamander is able to leave the water and hunt on land. Unlike fish, which have a two-chambered heart, all amphibians have a three-chambered heart. This allows that the animal's cells get more oxygen, since the animal expends more energy on land than it does in the water because of the loss of buoyancy.

Fertilization and development of the larvae vary by species. Some orders, such as the salamanders and legless caecilians, fertilize internally and lay eggs later. Others, such as frogs and toads, fertilize externally, sometimes in large groups. Some species of frogs lay their eggs in leaves that are full of rainwater. The larvae slide out of the leaves into rivers or ponds when they are slightly older and better able to defend themselves. Most amphibians are p animals, but some pare their reproduction down to just a few tadpoles, which the mother carries with her until the young adult animals are ready to survive on their own.

Typical tree frog

In the adult stage, all amphibians are insectivores. They eat insects, as well as spiders and tiny crustaceans. Some have remarkable adaptations for capturing prey, including, in certain frogs, a long, sticky tongue that acts as a "flypaper." Most newt and salamander species eat tiny worms or the larvae of insects, such as maggots.

There are three major orders of amphibians. The order Anura contains the frogs and toads, the order Urodela contains the salamanders, and the order Cymnophiona contains the worm-like caecilians. Amphibians are widely adapted to the environments in which they live. Some species are brightly colored, others exude toxins in their skin to protect from predation, while still others developed protective coloration and are rarely seen.

> **Amphibians are slowly disappearing from Earth.**

Amphibians are slowly disappearing from Earth. Part of the reason is destruction of their habitats, including wetland loss, but another major concern are toxins that humans flush into water systems, or spray on agricultural plants. Because of the amphibian method of respiration, they are subject to toxins in both water and air. Many deformed frogs and salamanders are being discovered in heavily polluted areas, some with horrifying birth defects, such as missing or extra limbs. Amphibians are perhaps more sensitive to these toxins than other organisms, but it is still a matter for concern, not only for the amphibians and those who rely upon them in food chains, but also for any other organism in the same ecosystem. Amphibians are indicator species that demonstrate the health of the ecosystem graphically.

# Exploration Activities

1. Why did the first vertebrates leave the water to live on land? What were the advantages and challenges of a terrestrial life?

2. Describe the amphibian's metamorphic life cycle.

3. Name the two ways in which amphibians are dependent upon water.

# The Amniotic Egg

The evolution of the **amniotic egg** drove the remainder of vertebrate evolution to date. Amniotic eggs enclose the embryo in amniotic fluid and provide a food source in yolk or by placenta. In the case of reptiles, birds, and a few mammals, the egg surrounds the embryo and food source with several membranes and a tough shell that protects the embryo from dehydration on land.

The amniotic egg made it possible for vertebrates to leave the water altogether, and return only to drink. *Amniotes* were thus able to colonize deserts, dry uplands, and grasslands—in short, to move away from the pond and into the rest of the world. Amniotes include reptiles, birds, and mammals, although the latter have modified amniotic sacs for the most part.

Amniotic eggs are all internally fertilized, as the shell is a barrier for sperm as well as for sand, dust, wind, and water. This changed the reproductive strategy of the vertebrates to an increasing K strategy, since parents invest time, energy, and resources to construct and defend amniotic eggs. More amniote offspring survive to reproductive age than non-amniotes.

The embryo is surrounded by the *amnion,* a membrane filled with fluid that cushions the small embryo and prevents its dehydration. In most higher mammals, the amnion exists within the uterus of the mother animal until the fetus is ready to emerge. Most amniotes are fed by a *yolk,* attached to the gut of the embryo, which provides proteins to the developing organism, and *albumen,* which is an additional food source and additional cushion for the embryo. The embryo develops and excretes wastes, which are processed by the *allantois,* a sac also connected to the gut of the embryo.

In placental mammals, food is provided and wastes are excreted by the *placenta,* an organ that takes food from the mother's body and excretes through her kidneys and other organs. In shelled amniotes, the allantois is left behind in the shell; in mammals, the placenta is expelled after the young. The mother often eats this nutrient-rich organ, which is a tell-tale sign of the birth of young to predators, but is also an important source of nutrition for a female who is nursing offspring.

Around all these organs is a membrane known as a *chorion,* which assists the allantois in gas exchange. Finally, in most reptile and all birds species, a hard shell protects the developing embryo.

The hard shell made it possible for eggs to be laid and covered with sand or forest litter, and to bear the weight of parent birds as they keep the eggs warm. Amniotes, although they may have a juvenile stage, do not go through metamorphosis. They are hatched, or born, with the same body form as adults. These innovations made vertebrate life far from oceans or ponds possible.

> **Amniotes, although they may have a juvenile stage, do not go through metamorphosis.**

 **Exploration Activity**

In the medium of your choice, create a model or drawing of an amniotic egg. You may need to do additional research to place all the parts in the correct location. Label the various organs and membranes.

## Reptiles

Snakes, turtles, alligators, and lizards make up the class Reptilia. Reptiles were the first class of vertebrates to benefit from the amniotic egg and move away from water sources. Unlike the amphibians with moist, rubbery skin, reptiles evolved a tough, scaly skin that prevents the loss of the body's moisture and gives some protection from predators. Unlike fish scales, which are made of bony or cartilaginous material, the scales of the reptiles are part of the skin of the animal.

Because they do not absorb oxygen through their skin, reptiles are entirely dependent upon a well-developed set of lungs. Like amphibians, most reptiles have a three-chambered heart. Crocodilians, however, have a four-chambered heart, like the birds and mammals. The four-chambered heart of the crocodilians completely separates the blood that has been oxygenated from that which has not been oxygenated. This gives crocodiles and their relatives the potential for great energy when needed.

There were a few other important changes that occurred between amphibian life and reptilian life. First, and most important, was the development of a skeleton in which the legs of the organism could hold the animal's body off the ground. This gave reptiles protection from hot sands and rocks, as well as the ability to move much more quickly than amphibians can move. Reptiles can escape predators or capture prey much more easily based on this simple change to the leg structure.

Other changes included the addition of claws at the end of the animal's feet. Claws assist in obtaining food, in defense, and in mating. Most amphibians do not have teeth, but most reptiles do, including specially adapted fangs, in some species, for injecting venom into prey animals. Many reptiles can unhinge their jaw in order to swallow prey animals whole. Finally, reptiles have defenses against wind and sand. They can close their eyes—a first in the animal kingdom—with specially adapted eyelids. They can shut their nostrils as well, if necessary.

Reptiles are ectothermic and must bask in sunlight for hours each day to adjust their temperature. When the temperature gets too warm outdoors, they slip into the shade or into a pond to cool down. Such behavior to change body temperature is called behavioral *thermoregulation*. For animals such as mammals and birds, thermoregulation is an automatic response. Because of the need to thermoregulate behaviorally, reptiles do not live in very cold climates. They prefer tropical climates, which is where the largest animals of their class occur, but many live in temperate zones and hibernate through the winter. Others, such as the sea turtles, live in the ocean, where temperature changes are not typically a serious threat.

> **Because of the need to behaviorally thermoregulate, reptiles do not live in very cold climates.**

## Obtaining Food

Because they are ectothermic, the metabolism of reptiles is much slower than the metabolism of birds or mammals. Consequently, they do not need to eat as much. A snake can go for several months after consuming a mouse or rat. They spend most of their long lives in a quiescent state between feedings.

Not all reptiles have the long stretch between meals. Turtles and tortoises, which are mainly vegetarian, graze on a more or less daily basis. Some turtle species are carnivorous, but with one notable exception, the snapping turtle, they seem to be content with food that doesn't make much of an effort to escape—snails and worms. Snapping turtles are very aggressive and will attack fish, amphibians, and small birds.

Most lizard species are insectivores. There are a few exceptions, notably the Gila monster and the Komodo dragon, which have been known to attack humans and other large mammals, and the Galápagos marine iguana, which eats algae.

Snakes are mostly carnivorous, but have several strategies for obtaining meals. Some, like the vipers, have venom that is injected into the prey through fangs, which operate like hypodermic syringes. The snake can fold the fangs up into its jaw when not in use to kill or subdue prey. Some snakes can spit poison into the eyes of a prey creature or potential predator and render it defenseless. Others, the constrictors, wrap themselves around a prey animal and squeeze the animal until it can no longer breathe. Still others capture crickets and other insects by suddenly popping out of a hole in the ground and swallowing the insect whole. All snakes swallow their prey whole.

Some snakes eat the eggs of other animals. Some eat grains and other vegetable-based foods.

Crocodilians are carnivorous and will eat anything that they can catch. They are very fast in the water, and only a little slower out of it. They typically eat fish, amphibians, small reptiles, and small mammals, but large crocodiles are known to take on larger mammals, such as antelope and even young hippopotamuses. Because of the loss of their habitat and the fact that they are now protected, alligators in the southeastern United States are moving into human areas. It is common for a Florida resident to find an alligator in his backyard! Many pets and other domestic animals are lost to crocodilians in this fashion, and in some cases, small children have been attacked. When humans encounter crocodilians, crocodilians are usually the winners. Before entering a tropical pond, lake, or river, even to wade, get information about whether the area has alligators living in it.

## Sensory Organs

Reptiles have extraordinary sensory organs. Snakes, for instance, have a specialized organ for "smelling" pheromones in the air. The organ is known as *Jacobson's organ,* and it is located in the roof of the snake's mouth. The snake uses its tongue to collect samples for this organ to analyze. Snakes can also "see" in the infrared frequencies, using a specialized organ on their heads. This allows predators to find their prey.

All reptiles have good vision, not unlike our own. You might notice that some species, especially predatory ones, have eyes on the fronts of their heads, while others, such as the turtles, have eyes on the sides of their heads. There is a biologically sound reason for this. Prey animals tend to have eyes on the sides of their heads so that they have vision that is as close to 360° as possible. Predators, on the other hand, don't need that. Instead, their vision allows them to judge how far away a prey animal is from them. Having eyes in front of the head allows the animal to have *stereoscopic* vision, and thus, depth perception.

## Reproduction

A few species of sea turtles migrate from their feeding grounds in the oceans to the beach where they were born. The mother lays her eggs, buries them in the sand, and leaves them alone. She swims back out to sea. Weeks later, the babies hatch and head for the water. Some make it, but many others are gobbled up by hungry seabirds. When they reach maturity, the babies who made it to the water will return to the same beach to lay their eggs. Other turtles and tortoises lay eggs in groups called *clutches,* which are guarded by one or both parents.

Alligator and crocodile mothers actually look after their nests until the babies hatch. She then carries them, one at a time, down to the water's edge. Depending on the temperature of the eggs, an alligator mother produces a clutch of all female or all male offspring. She can build a nest to produce the gender she wants.

Most snakes and most lizards lay clutches of eggs in protected areas and leave the offspring to chance. Some species, however, produce live young. Garter snakes, for instance, keep their eggs within their bodies. When the young hatch, the mother reabsorbs the shell material, and the babies are born alive. Garter snake mothers are protective of their young. They keep their young with them until they are nearly grown, and if danger threatens, the young hide in the mother's toothless mouth.

> **Most snakes and most lizards lay clutches of eggs in protected areas and leave the offspring to chance.**

 **Exploration Activities**

1. Choose one adaptation of early reptiles and explain how it allowed them to live on land.

2. What does reptilian ectothermy mean for the life cycle and survival of the animal?

3. Name two ways that reptiles care for their young.

## Birds

Birds, the cass Aves, are descended from the reptiles. There are 13 extant orders and currently around 9,000 species of modern birds. Biologists sometimes call birds "modern dinosaurs," because it is believed that the first birds were directly descended from small theropod dinosaurs about 100 million years ago. Some of the similarities between birds and reptiles can still be seen.

Birds have feet with claws, like the reptiles, and their featherless legs have the same sort of scales found on reptiles. Feathers themselves are modified scales. Fertilization takes place internally, and both birds and reptiles lay hard-shelled amniotic eggs.

Clearly, there are some differences. Birds can fly, a feat based on three characteristics that are different from reptiles and, in some cases, unique to birds. First, birds are the only class of organisms to have feathers. Feathers are essential for flight. As mentioned, feathers are modified scales, but they provide insulation and are lightweight enough that they enable flight. Feathers streamline the bird's body, decreasing wind resistance in flight. They also keep the animal warm at high altitudes. Birds must *preen*—they must remove dirt and make sure the flight feathers are in proper condition. Also, they must add oil to their feathers to keep them waterproof. Every bird has an oil gland near the base of its tail. The bird stimulates the gland with its beak and runs its beak all over the outside of the feathers.

There are three main types of feathers on a bird's body. Flight feathers are tough, scaly objects with a *rachis*—a long shaft up the center of the feather—as well as filaments that are held together by tiny hooks. When the hooks are properly held together, the animal is streamlined for flight. Softer, and closer to the body, are the contour feathers, which are shorter than flight feathers but also possess the hooked filaments. Finally, right next to the body, are down feathers, which have no hooks. The purpose of the down is to retain the animal's body heat during flight and at other times.

The second characteristic that aids flight is the fact that birds, unlike reptiles, are *endothermic*. Birds, together with mammals, regulate their body temperature through a very rapid metabolism. They can pant to reduce the amount of heat in their bodies. Also

> Birds, the class Aves, are descended from the reptiles.

> Birds regulate their body temperature through a very rapid metabolism.

like mammals, birds have a four-chambered heart. The amount of oxygen available to a bird is important for rapid metabolism. The bird's metabolism supplies the energy needed for sustained flight, during which the bird may beat its wings hundreds of times per minute. A powerful *sternum,* or breastbone, is the bone to which the flight muscles are attached.

American wetland birds

Birds have hollow bones. They have horny beaks, rather than bony teeth. If you have ever held a bird in your hand, you were probably struck by how light the animal is. Low weight is an obvious advantage to flight.

Since birds are endothermic, they are not restricted by temperature for their habitats. Birds live in Antarctica, as well as in steamy tropical rainforests. But endothermy, and a rapid metabolism, means that birds must consume large quantities of food to maintain their body temperature.

Bird reproduction nearly always begins with courtship. Males often change color, grow elaborate tail feathers, build bowers, or perform a complex song or dance to encourage a female to mate with him.

Parent birds build nests in trees or on the ground. They lay clutches of eggs, which they keep warm with their own body heat. The shape of eggs depends on where the egg is going to be laid. Sea birds, such as the brown pelican, lay eggs on exposed cliff surfaces. The egg is strongly pointed at one end, so when the egg rolls, it rolls around in a circle, rather than off the cliff. Other birds have evolved other adaptations. When the young hatch, they are covered only with down feathers, and cannot yet fly. Parents must spend a good percentage of the year hunting or gathering and feeding the always-hungry chicks.

> **Some birds, the raptors, are exclusively carnivorous.**

Some birds, the raptors, are exclusively carnivorous. They eat small terrestrial animals, mainly rodents and small reptiles. Some eat fish and mollusks in the ocean, marshy wetlands, inland lakes, lagoons, and ponds. Other birds are insectivores. Some eat water plants, while others eat fruits and seeds. Still others are nectar drinkers.

 **Exploration Activities**

1. Why do some biologists call birds "modern dinosaurs"?

2. What are the three adaptations for flight in birds?

3. Why do birds go through a mating ritual? How does this improve the quality of the offspring?

# Student Lab: Feathers

Birds have three main types of feathers: flight, contour, and down. In the preceding section on birds, the three types of feathers are described.

The structure of all feathers is basically the same—a long hollow shaft called a rachis runs down the center of the feather, and a series of filaments called barbs are attached to the rachis on either side. In contour and flight feathers, the barbs have hooked barbules at their ends. Barbules are clinging hooks along the sides of the feather structure that lock the feathers together so that air cannot pass through. Barbules do not exist in downy feathers.

In this lab, we will look at feathers and determine their type.

## Materials

- Microscope or hand lens
- Various bird feathers of all types

## Procedure

Under a microscope or a powerful hand lens, observe the structures of different types of feathers. Identify flight, contour, and down feathers. Find the barbs and barbules. Run your fingers against the grain of the feather to unhook the barbules, then smooth it back and observe the hooking action.

## Conclusion

Which feathers in your collection are flight feathers? Which are contour? Which are down? How could you tell?

## Mammals

Mammals, like birds, are endothermic, which means they can live in both very cold and hot climates. An entire order of mammals—the cetaceans—lives in the oceans exclusively. There are about 5,000 species of mammals alive today, and they all have a few important characteristics in common.

Every mammal, at some point in its development, has hair. In the case of the sea mammals, most of the hair is gone, but it covers babies at birth and remains in the form of whiskers even in these cases. Mammals shed their hair from time to time. You have probably seen fur come off your pet cat or dog, especially in the summer. With few exceptions, mammals shed a few hairs at a time. Mammalian hair serves several functions, including insulation, thermoregulation, protective coloration, and direct protection. Certain specialized hairs are exquisite sensory organs.

Mammals do live in hot climates, and often have to cool down. There are several methods by which mammals cool their bodies. First, mammals often have sweat glands, located in more or less hairless locations, that cool the animal down by secreting water onto the surface of the skin. As the water evaporates, heat is transferred from the body of the animal to the outside air. Another way mammals cool themselves is by rapid breathing, or panting. Panting releases water from the lungs, which results in cooling. Some mammals thermoregulate behaviorally. They cover themselves in mud, dust, or water.

Female mammals produce milk in mammary glands after the birth of offspring. Mammary glands are located on the ventral side of the female's body, in rows of two, with a total number somewhere between 2 and 12. Mammals make an incredible investment in each offspring, since they are K strategy animals. The act of milk production is often detrimental to the mother. However, the majority of any individual female's offspring survive, which means that the male (usually not part of the picture by birth) also has his genes passed along to the next generation. The strategy is a successful one for all concerned.

> Every mammal, at some point in its development, has hair.

Another characteristic all mammals possess is three middle-ear bones, which allow mammals to hear sounds from the outside world. These three bones, the *malleus,* the *incus,* and the *stapes,* provided a useful adaptation to prey animals and predators alike. Indeed, acute hearing is often all that saves an antelope from a lion. Two of the bones—the malleus and incus—come from the evolution of the mammalian jawbone.

Mammals have a diaphragm, a sheet of muscles located beneath the lungs in most species. The diaphragm separates the chest cavity from the abdominal cavity, where other organs and tissues are located. The diaphragm expands and contracts, facilitating mammalian respiration. Mammals, like birds, have a four-chambered heart that pumps enough oxygen through the blood to allow the animal to move very quickly.

## Diet and Adaptations

Most adult mammals have teeth, like most reptiles. A few, such as the anteater and platypus, never possess teeth. A biologist can look at one mammalian tooth and determine that mammal's diet. There are several kinds of teeth that are specially adapted to chew different sorts of food. Omnivores, like humans, have variations of all of them.

*Incisors* are the dominant teeth in rodents. They are the two primary teeth in the front of the mouth and are used for biting and gnawing. *Canines* are found in animals who eat meat. They are used for puncturing skin and tearing off flesh. *Molars* are the flat, wide teeth found at the back of the mouth. They are used to grind foods, often grains and grasses, and to crush food to make digestion easier. Grazers, such as antelope and cows, use these teeth most often.

Some grazing animals have multiple stomachs for digestion of cellulose-rich grasses. They regurgitate some of the previously swallowed food to chew it again. This process is called cud chewing. Once the food is swallowed again, it continues on to the other stomachs.

Mammals have widely varied diets, from plants to insects to animal tissue. Some mammals are scavengers and eat dead or decaying animal matter.

> **Mammals have widely varied diets, from plants to insects to animal tissue.**

## Mammal Reproduction and Care of Offspring

Young mammals are often born with no teeth at all, or with small deciduous teeth called milk teeth. This is because they will nurse, and adult teeth could injure their mother. Most mammals lose the deciduous milk teeth in childhood and grow adult teeth, which enable them to eat the food their parents eat. Childhoods vary among mammal species. Some mothers nurse for a short period of time before the young are able to forage on their own and may have several litters per year; others reproduce once per year; still others may have one offspring every 6 to 10 years. In general, the larger the organism, the fewer offspring it produces in any given litter, and the more time the mother takes in bringing the youngster to adulthood. Most mammals are capable of reproduction only when the young have ceased nursing, and the mother goes into *estrus*—a temporary period of fertility. Most mammals have a very overt estrus—the female exhibits very specialized behavior, her body changes, including overt color changes, and she is receptive to males. Once the ova has been fertilized, she is no longer receptive—she is pregnant.

Some mammals have a hidden estrus, and the female is usually receptive, whether fertilization is possible or not. Dolphins, bononos (a species of great ape closely related to chimpanzees), and humans all have hidden estrus. These animals continue to engage in sexual activity, even when the female is pregnant, and most sexual activity does not contribute to a pregnancy. However, these species engage in sexual activity for a very important reason. Males and females of these species are engaging in a specific behavior called pair bonding. Birds also pair bond, sometimes for life. The males of these species spend a great deal of time interacting with and caring for the offspring of the union. This allows the species to produce offspring more frequently, since they have the assistance of two parents beyond the nursing stage. This special adaptation brought these species back from the brink of extreme K.

There are three specific reproductive strategies that have evolved in mammals. The first, and most primitive, belongs to the subclass of the monotremes. There are only three species of monotremes—two types of echidna, and the duck-billed platypus. Monotremes lay eggs, and they do not have specialized teats. The mammary glands of the monotreme secrete milk through the animal's fur. Monotremes are only found in Australia and New Zealand.

> **Monotremes lay eggs, and they do not have specialized teats.**

Kangaroo

*Marsupial* mammals are well known for their pouches, and include opossums, kangaroos, and koalas, among others. There are 260 species of marsupials, most of which live in Australia and New Zealand, but there are a few species living in North and South America as well. Marsupials have a two-step birth process. They give birth to an extremely underdeveloped fetus, which crawls up from the uterus into the mother's pouch and attaches to the teat. While technically born, if the fetus were to be removed from the mother at this stage, it would die. Many species of kangaroos can put the pregnancy, at this stage, into suspended animation of a kind, if environmental conditions are too difficult to support the life of a newborn. When the fetus is fully developed, it emerges from the pouch, but during its childhood, often returns to the mother's pouch for reassurance and protection.

*Placental* mammals bear live young, which are more or less fully developed. The placentals nourish the young through a specialized organ which is attached to the uterine wall, called the placenta. Food the mother eats and oxygen the mother breathes are passed along to the fetus through the placental barrier. Wastes produced by the fetus are likewise passed through the placental barrier and are carried away through the mother's bloodstream, where her organs eliminate them from her body along with her own wastes. When the fetus has reached a particular gestational age, which is different in each species, the mother goes into labor, and strong uterine contractions bring the fetus and the placenta out of the uterus through the mother's vagina.

> **Placental mammals bear live young, which are more or less fully developed.**

## Mammalian Learning and Behavior

Mammals are generally highly intelligent organisms. They have a complex nervous system and a substantial brain with a great deal of surface area and specialized lobes that regulate certain processes.

As you have probably noticed, you can teach a dog or cat a behavior by *conditioning*.

All mammals have a childhood, and during that childhood, their mother and, in some cases, father and extended family teach the offspring what they need to know about the world. Not all of the teaching is extensive, but even mammal parents who only spend a short time with their offspring, such as field mice, demonstrate to the offspring what sorts of seeds are safe to eat, where sources of water can be found, and where to hide when a predator comes along. Other mammal parents, who spend far more time with their offspring, teach many other things as well.

> **Most mammal behavior can be modified by conditioning.**

Most mammal behavior can be modified by conditioning. When you train your dog to sit, you help your dog associate the word "sit" with a reward of some kind—a scratch on the head, praise, or a small treat. The dog soon associates the pleasurable response, called *positive reinforcement,* with the performance of the trick, and repeats the behavior until it is firmly ingrained. What you are doing, although you may not be aware of it, is building on the animal's innate desire to belong to a group—a pack, in the case of dogs. Mammals, unlike other animals, are highly social. With few exceptions, they desire the company of others of their own kind, even when not breeding and rearing young. Usually mammals group into families of females and their young, with perhaps one male standing guard, but there are exceptions. Packs of wolves and others of the dog family are centered on a central breeding couple—the alpha male and alpha female. The rest of the pack assist the breeding couple with rearing the offspring and do not reproduce themselves. The pack hunts together, and feeds and cares for the young together.

A young mammal within any family grouping soon learns appropriate and inappropriate behavior, depending upon how the adult animals respond. Although some mammalian behaviors are instinctive, even among humans, the majority of behaviors are learned in youth. Mammals play to establish and solidify roles in their societies, unlike most other animals.

## Origins

The earliest mammals are believed to have sprung from a group of animals called *therapsids,* some 200 million years ago. Therapsids and the first reptiles share a common amphibian ancestor. Therapsids were well known for their ability to thermoregulate using large, fan-like plates on their backs. Until recently, therapsids were believed to be early dinosaurs. Today, it is believed that therapsids and dinosaurs share a common ancestor but are not directly related.

Mammals were small and inconspicuous throughout the Mesozoic era, while dinosaurs reigned. Only after the mass extinction of the dinosaurs, 65 million years ago, did mammals begin to acquire any size at all. Mammals were able to radiate into many niches, especially the new grassland niches that were made possible by the evolution of flowering plants. In the Cenozoic era, mammals became gigantic. Toward the end of our period, the Quaternary period, mammals began to acquire roughly the size they have today.

> **There are 19 orders of mammals.**

## Diversity

There are 19 orders of mammals. Two of the orders represent the monotremes and marsupials. The others are placentals. The largest living animal, the blue whale, is a member of the order Cetacea, which includes all the whales and dolphins. Cetaceans had ancestors who once lived on land, but returned to the oceans after the Mesozoic era. They breathe oxygen directly from the air, but remain under water for hours at a time. Order Primate includes all monkeys, lemurs, great apes, and humans. Other placental orders include Chiroptera, which includes bats; Rodentia, which includes mice, rats, squirrels, and beavers; Edentata, which includes anteaters; Lagomorpha, which includes rabbits and hares; Insectivora, which includes moles and shrews; Carnivora, which includes bears, the great cats, and wild dogs; Pinnipedia, which includes seals and walruses; Sirenia, which includes manatees; Proboscidea, which includes the elephants; Perissodactalya, which includes all hoofed mammals with an odd number of toes, including horses; and Artiodactyla, which includes all hoofed mammals with an even number of toes, including cows and hippopotamuses. Four small orders have only one or two members each. These are the orders that contain aardvarks, hyraxes, gliding lemurs, and horned anteaters, such as the armadillo.

 **Exploration Activities**

1. Compare the reproductive strategies of monotremes, marsupials, and placentals.

2. You discover a fossilized mammal skeleton dating from the early Cenozoic era. The skeleton is about the size of a dog, but has hooves, three toes, and flat teeth. What do you infer about the animal based on what you know about mammals?

3. Design an experiment to teach a young rat to turn right consistently in a maze. If you have a pet rat, try the experiment. What are your conclusions?

**BACKGROUND**

# Defining Evolution

What is evolution? Evolution is a change in the gene pool of a population over a period of time. Genes, you will recall, are the hereditary units carried by parent organisms and passed to their offspring at reproduction. The gene pool is the grouping of all genes in the entire population.

Driving evolution are two important factors—*natural selection* and *sexual selection*.

Let's consider a modern example of evolution, which has been documented since the 1800s. In England there is a species of moth known as the peppered moth, which is known to rest on trees covered with lichen. The moth, in the mid 1800s, was a gray insect with specks of black, and on lichen-covered trees it was nearly invisible. Its protective coloration saved it from its many predators. Occasionally, due to a recessive trait, a solid black insect would emerge from the cocoon. However, the black insect was spotted by birds almost immediately, and few survived to pass on their genes.

The coloration of the moth is determined by a single gene. Before 1848, the black moths made up less than 2% of the population of the species.

Times changed, and heavy industry in the area spewed pollution into the air, which killed much of the lichen. The trees became a uniform dark color. The speckled moths could be spotted easily, and were gobbled up by the birds, while the black version often went unnoticed. As a result, more of the black moths reproduced, and the numbers of black moths versus speckled moths reversed. By 1898, 95% of the peppered moths in highly industrial areas were black. Now that environmental conditions are improving and the lichens are growing back, the situation is changing again.

What drove the moths' remarkable evolution was natural selection. In natural selection, animals' natural processes and defenses are either selected for or selected against, depending upon conditions in the natural world. This causes animals to either move to a region where their natural defenses are more useful—some moths left the heavily polluted city areas for the countryside where their

> Evolution is a change in the gene pool of a population over a period of time.

defenses still worked; *adapt*—the population in the heavily polluted areas produced more black moths; or simply *die out*—if there had been no recessive black moth, the species would have become extinct in polluted areas.

The other major factor in higher organisms—sexual selection—can be seen in the tail feathers of the peacock. The ornate tail feathers are not useful for flight, nor do they have any functional purpose, save one. Peacocks use their tail feathers to attract females during mating rituals. Peahens are attracted to the male with the biggest and most beautiful feathers. They will choose to mate with those peacocks. The offspring produced will likely also have long, beautiful tail feathers, and will have further opportunities to mate than those without the long feathers. Sexual selection belongs to the female of the species, who makes the choice about what is considered a useful attribute. If foxes, or other natural predators, began to be attracted by long peacock feathers as well, the females would begin to see those feathers as a liability for their chicks, rather than a desirable attribute, and they would soon disappear from the population.

> **Evolution does not happen in individuals.**

Evolution does not happen in individuals. Each organism is locked into his or her own genetic structure for life. Evolution only happens to populations and, usually, over a long period of time, because it takes many generations for a characteristic to either manifest itself or die out of a population of organisms through selection.

All kingdoms of life experience evolution. In some, especially in prokaryotes and other organisms that reproduce asexually, the sole mechanism for evolution is natural selection. Since the young are usually carbon copies of the adults, the chance of mass extinction among these organisms is very high. The only way evolution can occur is if a copy error occurs during cell division. Although infrequent, copy errors have had enormous impact on Earth's life forms and environment. One extremely important copy error gave rise to photosynthesizing bacteria; another gave rise to eukaryotic cells.

##  Exploration Activities

Consider each of the following evolutionary conditions, and determine whether the method that drives the condition is sexual selection or natural selection.

1. Two formerly identical species of finch live on two small neighboring islands in the Pacific. One island has a burrowing beetle on it that does not live on the other island. The finch on the island with the beetle has developed a twisted beak, which is capable of holding the slippery insect, and pulling it from its burrow. The other finch has a straight beak for eating mosquitoes.

2. The females of a certain species of seal rest on ice floes during mating season. One by one, male seals visit them, bringing fish. The females choose a male who brings the largest fish to the ice floe, and all the females mate with that male.

3. A type of African beetle creates a large ball of animal dung. The ball of dung remains warm during the rainy season. The male creates the nest, then waits for a female to choose that nest. Once a female chooses the nest, the pair mate.

4. In the rainforest, where light on the ground is dim, certain plants have developed large, fan-like leaves. The leaves of these plants capture more sunlight than normal leaves, and photosynthesis can occur.

# Student Lab: Creating "Life"

Scientists estimate that life on Earth originated between 3.9 and 3.5 billion years ago. Life most likely arose from chemicals present on early Earth. Carbon-based life, the only variety we are familiar with, is based on a number of amino acids, which are present in comets. During the Great Bombardment as the planet was forming, the amino acids were probably transferred to Earth along with the water they carried. Amino acids arise from other chemical reactions rather easily, both in extremely cold and extremely hot conditions, such as would have existed on the primitive Earth without a moderating atmosphere. One of the most important steps in early evolution was the enclosure of organic compounds by a membrane, creating a *protocell*.

## Materials

- Microscope, slide, and cover slip

- Labware, including 50-milliliter (mL) graduated cylinder, 50-mL Erlenmeyer flasks (2), and a 400-mL beaker. Heat-resistant glass, such as Pyrex cups, can be substituted.

- Gram scale

- 1% NaCl solution

- The following amino acids: aspartic acid, glutamic acid, and glycine. None is dangerous, but all are smelly—don't spill them on yourself!

- Hot plate or kitchen stove

- Ring stand and clamp

- Tongs

- Watch or clock with second hand

- Stirring rod

- Pipette

- Apron, oven mitts, and goggles

**Special Safety Consideration:** Use caution when operating the hot plate.

## Procedure

1. Pour 250 mL of water into the beaker, and place the beaker on a hot plate or in a hot-water bath on a stove.

2. Clamp one of the flasks to the ring stand.

3. Add to the flask 0.5 grams of each of the following: glycine, aspartic acid, and glutamic acid. Stir thoroughly.

4. When the water in the beaker begins to boil, loosen the clamp and lower the flask into the beaker.

5. Heat the amino acids for 20 minutes, keeping the water in the beaker to a simmer.

6. Measure and pour 5 mL of NaCl into the second flask, and put the flask on the hot plate.

7. When the NaCl solution starts to boil (this will happen very quickly), use the tongs to remove it. Slowly add the NaCl solution to the amino acid solution. Turn off the hot plate.

8. Raise the flask containing the amino acid solution, and allow the contents to cool for 10 minutes.

9. Prepare a wet mount using one drop of the mixture, and observe under low power. Locate and diagram a protocell.

10. Switch to high power to examine the protocells.

## Conclusion

After you watch for a while, you will note that your protocells are growing. This is because they are metabolizing the NaCl. Soon, you will observe them dividing. They are not true life, since they do not possess DNA, but these simple cells were most likely the precursor to true life.

 **BACKGROUND**

# Precambrian Time

Earth was formed along with the Sun and the rest of the solar system about 4.6 billion years ago, in a series of collisions of asteroids and comets called the Great Bombardment. The Bombardment caused the young Earth to heat up by friction, and it was nearly half a billion years later before the Earth was cool enough to allow the formation of liquid water. All life forms we are aware of require the presence of water in its liquid form, so this was an important step in the history of life. This earliest period of time on Earth is known as the Precambrian time. Until recently, it was believed that the first life originated in the Cambrian period of the Paleozoic era. Now, it is known that the earliest life forms actually arose during the Precambrian time, the time on Earth prior to the Cambrian period.

As the oceans formed, between 3.9 and 3.5 billion years ago, a series of complex chemical reactions took place which eventually led to the formation of extremely simple life forms—the prokaryotes. The first prokaryotes, the archaebacteria, could not tolerate the presence of oxygen, but around 3.5 billion years ago, the first photosynthetic organisms arose—the cyanobacteria. They were responsible for the first great mass extinction on Earth. Most anaerobic archaebacteria died out. Six more mass extinctions would follow.

If you consider the entire history of Earth as a calendar, the formation of Earth occurred on the first of January, and the first prokaryotes did not arise until the middle of March. But arise they did, and we know of their early existence because they left a fossil record. In the deserts of Australia, and elsewhere on the planet, strange, bulbous rock formations called *stromatolites* can be found. Stromatolites were formed from thick mats of cyanobacteria, whose sticky filaments collected rock and sand particles, and precipitated *calcium carbonate*, preserving the fossil of the bacteria. Over time, the structures grew, layering generations of cyanobacteria upon one another. The oldest part of the structures dates from about 3.5 billion years ago.

Almost all of the life forms throughout the Precambrian time were unicellular and prokaryotic. However, toward the end of the era,

> **Almost all of the life forms throughout the Precambrian time were unicellular and prokaryotic.**

> Precambrian time ended as the Paleozoic era began—around 545 million years ago.

some eukaryotic cells arose, and finally, simple multicellular organisms appear in the fossil record. Like the cyanobacteria, the new organisms were soft-bodied and left only imprinted fossils. They included animals such as simple sponges and other *parazoic* organisms, algae, jellies, and protozoan protoctists.

By our hypothetical calendar, the Precambrian time did not end until the middle of October. In real time, Precambrian time ended as the Paleozoic era began—around 545 million years ago.

 **Exploration Activities**

1. Which organisms caused the world's first great mass extinction? How? Which organisms died out as a result?

2. What is a stromatolite, and what is its importance in pinpointing a date for the origin of life?

3. How long did Precambrian time last? Why did it take so long for life to originate? What kinds of organisms were present at the very end of Precambrian time?

# The Paleozoic Era

The Paleozoic era lasted from 545 to 248 million years ago. The name Paleozoic means "old life," and the era is characterized by a vast increase in the number of phyla of all kingdoms of life.

Physically, the Earth had changed. Earth's core had cooled, and the oceans were well established. Temperatures were not greatly different from modern-day temperatures. The volcanic eruptions that dominated the Precambrian had largely ended, and few meteor impacts occurred. The composition of the atmosphere had changed, owing to the amount of oxygen that had been released into the air by the algae and cyanobacteria.

The oceans had a large amount of dissolved calcium carbonate in them, and this could have been dangerous to small animals and some algae. However, during the Paleozoic era, a good deal of this dissolved calcium carbonate became bound up into the hard body parts of living organisms. The hard shells, coral deposits, and later, bones of animals were often preserved, and give us a very clear idea of what kinds of life existed at various times during the Paleozoic era.

By the end of Precambrian time, there were hundreds of species of bacteria, along with many protoctist species, and a few animal species. No plants or fungi yet existed. All organisms had soft bodies and were only sporadically preserved. By the end of the Paleozoic era, all phyla currently represented on Earth were in existence, and many of today's orders were also present.

The Paleozoic is divided into six major periods. The first is known as the Cambrian period. Scientists often refer to a Cambrian "explosion" of life that started at the beginning of the Paleozoic era. During this period, the shallow seas of the Earth were full of many types of invertebrates, including worms, mollusks, echinoderms, and even primitive arthropods known as trilobites. The diversity of the period is captured in shale deposits around the world, including the remarkable Burgess Shale of Canada. The Cambrian is often referred to as the Age of Invertebrates. No terrestrial plants yet existed, and no chordates were yet in place.

> The Paleozoic era lasted from 545 to 248 million years ago.

The next period was the Ordovician period. During this time, invertebrates in the oceans diversified. Trilobites reached their ultimate sizes, corals and crinoids arose, and all mollusk types became represented, including cephalopods, bivalves, and gastropods. Algae also diversified, with all three types of multicellular algae represented, as well as diatoms. The ancestor of the sea squirt lived during this period, and the first vertebrate—primitive fish—began to appear in the fossil record. There are fossil spores that suggest that plants may have begun to colonize the land around this time.

The end of the Ordovician period was marked by a mass extinction and an ice age, and the climate changes caused great upheaval. The next period was the Silurian period, which saw a stabilization of Earth's climate. Among other things, the continental ice sheets melted, increasing the ocean depth. There were several large changes to living systems, as well. The corals had begun to live in colonies, known as coral reefs, and fish diversified. In addition to the primitive jawless fish, new animals with jaws emerged, and fish also colonized freshwater environments. In fact, the Silurian period is often called the Age of the Fishes. On land, the first arthropods had emerged, and grown to monstrous proportions, while the first vascular plants, such as ferns, were fossilized.

By the next period, the Devonian period, life on land was well-established. In addition to a wide diversification of terrestrial invertebrates, the first terrestrial vertebrates also appeared. Most vegetation during the early period was small, but by the end of the Devonian, seed plants had emerged, and the first forests were established. Back in the seas, fish continued to diversify, and invertebrates also continued to branch out.

The next period was the Carboniferous period. Carboniferous means "coal bearing"; much of the fossil fuels used today had their origins in this period. The climate had changed dramatically. During the late Devonian and through the Carboniferous periods, the temperature became warm and humidity achieved steam-bath proportions. Plants skyrocketed in size and diversity. The large land invertebrates shrank in size, in response to the threat of the new vertebrates. Not only were amphibians on land, but amniotes had also emerged. The ancestors of birds, mammals, and reptiles were able to reproduce on land without returning to the pond, which increased their range. In the oceans, fish ruled the seas.

Sharks and bony fish increased the range of vertebrates in the water as well.

The last period of the Paleozoic era was the Permian period. It is most famous for its end.

> **The PT boundary marked the largest mass extinction in history.**

The PT boundary, named for the Permian/Triassic periods, marked the largest mass extinction in history, larger, even, than the one that would kill off the dinosaurs almost 200 million years later. Marine communities were the hardest hit. Most of the marine invertebrates then alive, including the crinoids and trilobites, were wiped out. On land, a smaller extinction killed many *diapsids* and *synapsids,* vertebrate animals that were the ancestors of mammals and dinosaurs. A shift from mostly spore-bearing to mostly seed-bearing plants occurred, and large conifers became the order of the day.

By the end of the Paleozoic, life had colonized every inhabitable region on Earth. The explosion of diversity that began the era was carried through to its end. The next great era would see the rise of dinosaurs and flowering plants, and Earth would never be the same again.

 **Exploration Activity**

Draw a line from each period name to an important fact about that period.

Cambrian          first terrestrial plants appear

Ordovician        Age of the Fishes

Silurian          fossil fuels laid down

Devonian          ended with mass extinction that wiped out trilobites

Carboniferous     first forests appear

Permian           "explosion" that represents most modern phyla

*Top Shelf Science: Biology*        **EVOLUTION**

## The Mesozoic Era

The Mesozoic era began around 248 million years ago and ended abruptly 65 million years ago. On our calendar, the Mesozoic era begins on December 10.

Physically, the Earth made some of its most important changes, at least on land. The large supercontinent of Pangaea broke up during the Mesozoic era, and continents began to drift apart. The movement of the plates caused earthquakes, volcanoes, and tidal waves, and built large mountain chains. The world was warmer. As far as science can tell, there were no ice caps at the North and South Poles at the beginning of the Mesozoic era.

There were three periods in the Mesozoic era. The Triassic period, beginning around 248 million years ago, was the period in which mammals made their first true appearance. Early mammals were very small and very inconspicuous. The large fern forests were dying out and being replaced with coniferous forests, but many spore-bearing plants remained, as well as other gymnosperms. Flowering plants had not yet appeared. In the oceans, fish continued to reign as the dominant form of life, but invertebrates were making a comeback after the devastation of the PT mass extinction.

The Mesozoic, or "middle life," era is also known as the Age of the Dinosaurs. The first dinosaurs appeared during the Triassic period. Some were quite large. The first known dinosaur fossils date from about 220 million years ago, and they are of the herbivore *Plateosaurus,* who was a giant of the time at about 5 meters in length. It had a large, curved thumb claw that was probably used for feeding. There were a few predatory dinosaurs around, as well, but they were not large and probably did not feed on animals like *Plateosaurus.* All dinosaurs from this period were *bipedal*—that is, they walked on two legs.

By the next period, the Jurassic, the humidity once again increased. Dinosaurs became larger, and for the first time some of them were quadrupeds. Large, bulky animals, such as *Apatosaurus,* roamed the swamps, eating soft swamp plants, while the water assisted in keeping the animal buoyant and able to move. *Apatosaurus* was a new sort of herbivore—a *sauropod,* an animal with legs like a

> The Mesozoic, or "middle life," era is also known as the Age of the Dinosaurs.

reptile tucked under the dinosaur, instead of out to the side. *Apatosaurus* could measure 35 meters from head to tail, and other sauropods were even bigger. Armor-plated dinosaurs also appeared, such as *Stegosaurus*. Large theropod predators emerged, able to tackle even the sauropod giants. *Allosaurus* was just a little smaller than *Tyrannosaurus,* but it was every bit as fierce.

In the late 1850s, a small skeleton was discovered in a limestone quarry in Germany. It was of a 1-meter-long, bipedal theropod called *Compsognathus*. It received much acclaim at the time, because it was fossilized right after its last meal, and the contents of its stomach were also perfectly preserved. Just a few years later, another *Compsognathus* skeleton was found at the same dig . . . or so the discoverer believed at first. A closer look, however, revealed something very unusual about this particular specimen—imprinted into the limestone were the clear impressions of feathers. This was no *Compsognathus*—it was the first bird, *Archaeopteryx*. *Archaeopteryx* is a transitional species. It possessed the teeth of the theropod it resembles, solid bones of a reptile, and a long, bony tail. But it was also unquestionably a bird. *Archaeopteryx* demonstrates as clearly as the lobefishes and lungfishes that populations of animals change over time—even across order, phylum, and kingdom boundaries.

> **Archaeopteryx is a transitional species.**

The last period of the Mesozoic era was the Cretaceous. A new, successful form of dinosaur, with a hip like a bird rather than a lizard, emerged at the end of the Jurassic, but covered the globe during the Cretaceous. They are known as orthinopods. They could move fast and eat by chewing leaves and green plants in the backs of their mouths, which meant that they could literally eat on the run. They probably had to do so, because their new neighbors were members of the Deinonychus family, best known for the *Velociraptor,* a theropod with a slashing claw. *Velociraptor* and its relatives hunted in packs, like modern wolves, and were very intelligent animals. Other animals evolved with defensive weapons on their bodies. *Triceratops,* for instance, had three long horns that it could use to gore a predator. *Ankylosaurus* wore armor plating and had a bony tail with a spike in it. *Tyrannosaurus rex* is probably the most famous of the dinosaurs from this period. It and its relatives were huge. *Tyrannosaurus* could stand 6 meters high and measured 15 meters from snout to tail.

> **A crater was recently discovered in the Yucatan Peninsula of Mexico that is the right age to be the dinosaur's doom.**

During the Cretaceous, mammals radiated, and while they still kept a low profile, they diversified to fill available niches. Flowering plants began to blossom all over Earth. Reptiles took to the air and to the sea.

But around 65 million years ago, the world changed again. Currently, it is believed that an asteroid or large meteor struck Earth near the Yucatan Peninsula. The impact was so great that dust and dirt were thrown up many kilometers into the air. Forests were set ablaze by the force of the impact. Sunlight was blocked for months by debris in the atmosphere, and plants, sauropods and orthinopods, then theropods finally died out from lack of food and suddenly colder temperatures. The evidence for such an impact is abundant. There is, around the world, a layer of an element called iridium in the fossil record precisely at the time when the dinosaurs died out. Iridium is not found naturally on Earth's surface, but it is found in meteors, asteroids, and comets. A crater was recently discovered in the Yucatan Peninsula of Mexico that is also the right age to be the dinosaur's doom. Whatever happened at that time, the dinosaurs, along with many other species, died out. Mammals and birds took advantage of the new conditions to colonize the world.

 **Exploration Activities**

1. What are the three periods of the Mesozoic era?

2. What was the defining life form of the Mesozoic era?

3. What other important life forms existed at this time?

4. What was the importance of the emergence of flowering plants, as far as mammal evolution is concerned?

5. What is the probable reason for the mass extinction of the dinosaurs 65 million years ago? What is the evidence for such an event?

**BACKGROUND**

> The Cenozoic era began 65 million years ago and continues to this day.

# The Cenozoic Era

The Cenozoic era began 65 million years ago and continues to this day. On our calendar, the Cenozoic era would have begun on December 26. The Cenozoic is sometimes referred to as the Age of Mammals, because during this modern period mammals truly flourished.

Physically, the Earth's continents were in the same approximate positions they are today. Some landforms were submerged, such as the Florida peninsula. Florida, and other landmasses, would be uncovered only later in the era, when Arctic and Antarctic ice bound a great deal of seawater into glaciers, and sea levels dropped worldwide. During the Cenozoic, plate tectonics played a major role in the shape of the continents. It was during the early Cenozoic, for instance, that the subcontinent of India joined with the larger Asian landmass, and the Himalayas were formed.

The Cenozoic era is divided into two major periods, the Tertiary and the Quaternary. By far, the Tertiary was the longest period, extending from the end of the Mesozoic until comparatively modern times, around 2 million years ago. The Tertiary is itself divided into five epochs.

During the first epoch, the Paleocene, mammals radiated rapidly. At the beginning, mammals formed two basic groups—the hoofed herbivores, and an early group of carnivorous animals called the *creodonts*. One species of creodont took to the water and spawned an entire order of mammals—the cetaceans. Most of the mammals did not resemble their modern descendents, even though the modern descendents still occupy the niches their ancestors filled 65 million years ago. This was the time of the rise of the grasslands, and new ecological niches formed by deciduous forests also opened up. In Australia, which existed in virtual isolation, marsupials and monotremes evolved on separate paths.

During the Eocene, the world's first primates, bats, and horses appeared. During this epoch, the giant mammals began to appear. One important one was *Baluchitherium,* a giraffe-like rhinoceros that measured 6 meters at the shoulder and was 9 meters long. This species was the largest land mammal that ever lived. The Eocene lasted from 55 million years ago to about 35 million years ago.

The next epoch was the Oligocene, which saw the first mastodons, as well as additional members of the rhinoceros family. Hoofed mammals continued to diversify, and some of them became giants. True rodents appeared. Angiosperms that had previously lived in swampy areas radiated into huge savannas. Grasslands might look soft and lush, but in fact, the grasses are abrasive. Silica and cellulose give the grasses their ability to stand upright, and eating silica is like eating shards of broken glass. During this epoch, the hoofed mammals evolved several strategies for breaking down this difficult food, including multiple stomachs, thick tooth enamel, and the ability to regenerate teeth lost in the battle against the grasses. The Oligocene lasted until about 24 million years ago.

During the Miocene, primates diversified into ape-like organisms. One of these organisms was the common ancestor of the modern great apes and humans. Grasslands and deciduous forests were the dominant ecosystems on the planet. The Miocene lasted until about 5 million years ago.

The last epoch in the Tertiary period was the Pliocene. The Pliocene saw the evolution of true *hominids,* which we will discuss in the next section. During the Pliocene, the weather was changing again from mild and dry to cold and icy. The Pliocene lasted until about 1.7 million years ago.

The Quaternary period, by comparison, has been very short, but it isn't over yet! There are two epochs in the Quaternary period.

The first, the Pleistocene, was a period of ice age. Mastodons, saber-toothed cats, and other familiar ice-age mammals were plentiful. Gigantism continued, including giant members of the sloth family. True human beings arose during this period. The Pleistocene epoch ended as historic times began, some 10,000 years ago.

The historical epoch, the Holocene, is the period in which all modern animals, and modern humans, appeared. The Holocene is well known, because during the whole of it, humans have recorded history. We continue to live within the Holocene epoch today.

> **The historical epoch, the Holocene, is the period in which all modern animals, and modern humans, appeared.**

 **Exploration Activity**

Draw a line from each epoch name to an important development that occurred within that epoch.

Paleocene          First rodents appear

Eocene             First apes appear

Oligocene          First giant mammals appear

Miocene            First humans appear

Pliocene           Major mammal radiation into many niches

Pleistocene        Epoch of historical records

Holocene           First hominids appear

## Hominids

The story of human evolution begins with hominids. Hominids were *bipedal* apes that began to appear in the fossil record around 5 million years ago. Hominids were different from the other great apes in a number of ways. They could walk upright, although many of them continued to walk on knuckles from time to time. They developed group living arrangements that included adult and juvenile males and females. As they evolved, additional changes to jaws, brain size, and foreheads differentiated the families.

The first suspected bipedal hominid was *Ardipithecus ramidus*. This animal looked very much like a small ape, and it was probably mainly arboreal. However, evidence suggests it was bipedal, making it a member of the hominid family.

*Australopithecus anamensis* is another early biped that lived between 4.2 million and 3.9 million years ago. The word Australopithecus means "southern ape."

Next, and most famous of the early hominids, came *Australopithecus afarensis*. This animal lived around 3 million to 4 million years ago and showed evidence of a human-like pelvis and leg bones. A young female specimen was discovered in 1976 and nicknamed Lucy. She stood about 1 meter tall. Additional members of her species were found the next year near the same dig, making *A. afarensis* one of the most well-understood of the early hominids.

*Australopithecus africanus* was very similar to *A. afarensis,* but had a larger body size and brain size. They lived between 3 million and 2 million years ago.

Next, *Australopithecus aethiopicus* appeared. They lived between 2.6 million and 2.3 million years ago and appear to be a transitional species. Some parts of the skeletons are quite advanced, while some show primitive characteristics.

Two other species, *Australopithecus robustus* and *Australopithecus boisei,* are most likely differently sized organisms of the same species. They were the first tool users in the family that would eventually include human beings. They had a larger skull than previous hominids, and more finely adapted teeth.

 **Exploration Activity**

Hominids are ancestors of humans. There are several possible "family trees" that lead from hominids to humans. Do some research on the findings of various paleoanthropologists, such as the Leakeys, Donald Johanson, and others, and create a family tree that includes hominids and true humans. There are some dead ends along the way!

## Human Evolution: *Homo*

The various species of *Homo,* the true humans, started to appear in the fossil record around 2 million years ago. The first true human is *Homo habilis,* the "handy" man. Like his late hominid ancestors, *H. habilis* made tools of bone, wood, and stone. Stone tools were a developmental breakthrough in the history of humanity and kicked off the Paleolithic (or old stone) Age. *H. habilis* was fairly tall and sturdy, and possessed a brain slightly larger than the hominid ancestors.

*H. habilis* had a good run. They virtually had the planet to themselves for almost a million years. The next species of human was *Homo erectus,* sometimes called Java Man, because remains were discovered on the Indonesian island of Java. Skeletal remains have been discovered in Africa and Asia, which suggests that this species migrated during its tenure. The *H. erectus* skeleton shows a trend toward becoming *gracile,* or thin boned. It looks remarkably like our own skeleton, except for the skull, which is still quite primitive. It is believed that *H. erectus* used fire as protection from predators.

*Homo heidelbergensis* was another step on the road to our own species. *H. heidelbergensis* mainly inhabited Europe, where they had spread from Africa. They had a larger brain than *H. erectus,* in a smaller skull. This was another gracile species, like our own.

*Homo neanderthalis,* on the other hand, which inhabited the same regions, was an evolutionary dead end. *H. neanderthalis* was a highly intelligent cousin with a *robust,* or heavy-boned, frame. They had larger brains than modern humans and built extraordinary stone tools. They also built musical instruments, buried their dead, and appeared to perform ritual sacrifices, signaling a nascent religious belief system. As interesting as the species was, they did not contribute to modern humans. Neanderthals died out around 28,000 years ago, without leaving descendents.

*Homo sapiens,* modern humans, have inhabited the planet for some 120,000 years, and have migrated to every possible location around the world. *H. sapiens* are gracile, with an extremely streamlined frame. Modern humans and Neanderthals shared Europe for a

> **It is currently believed that all species of *Homo* had their ancestral origins in Africa.**

time; it is possible that there was limited interbreeding. However, the typical features of modern-day humans are much more similar to early *H. sapiens* (sometimes called Cro-Magnon humans) than to *H. neanderthalis*. The brain of *H. sapiens* (whose name means "wise man") is large, intricate, and capable of abstract thought. *H. sapiens* used language. Early *H. sapiens* painted representational scenes in caves and on rocks. They domesticated animals, built shelters, increased the sophistication of stone working, built boats for water travel and fishing. Eventually, by the Neolithic (or new stone) Age, they showed a marked propensity for religious belief. Stonehenge and other standing stone monuments were built by Neolithic *Homo sapiens*. By 10,000 years ago, *H. sapiens* had developed sophisticated cultures, built buildings from stone, developed agriculture, and started the world's first cities.

It is currently believed that all species of *Homo* had their ancestral origins in Africa. Until comparatively recently, the Mediterranean basin was dry and could be crossed by foot. A radiation of *H. erectus* from Africa was followed, 200,000 years ago, by a second radiation of early *H. sapiens*. Other theories of human origins, such as of parallel evolution in South America and China, are still under investigation.

 **Exploration Activities**

1. Which human first used stone tools?

2. What is the difference between gracile and robust bone structure? Which do modern humans have?

3. A friend tells you that Neanderthals are ancestors of modern humans. How do you refute this notion?

# Answer Key

## What Is Biology?

1. kingdom, phylum, class, order, family, genus, species

2. Viruses do not perform all the functions of life, especially growth and independent reproduction.

3. Viruses reproduce by invading host cells, "reprogramming" the host cell's DNA with the virus's own RNA or DNA, replicating and creating new virus particles within the cell, and finally destroying the cell in order to release the new virus particles.

## What Is Ecology?

1. biosphere, biome, ecosystem, niche

2. Answers will vary, but an example is prairie ecosystem.

3. Abiotic factors are all nonliving factors; biotic factors are alive or were once living.

4. Flora means anything that photosynthesizes, as well as fungi; fauna is all animals and any consumer organism.

5. a species, usually near the top of the food chain, whose presence or absence gives biologists a clue about the health of the ecosystem

6. Diversity is the number of species within a given biome.

## Symbiosis

1. commensal
2. parasitic
3. mutual

## Food Chains

1. apple tree is producer, fruit-eating finch is first-order consumer, bobcat is second-order consumer, coyote is scavenger, bacteria is decomposer

2. to remove dead organic tissue, and to break down organic matter for use by producers

3. A food web has multiple consumers and producers.

4. Autotrophs are producers, and they make their own food. Heterotrophs are consumers.

## Larger Patterns in Ecosystems

1. Answers will vary; possible answers include: seasonal variability, weather or climate change, migration patterns

2. A short-term change often results in loss of generations, which decrease the populations for years.

3. Long-term changes can cause mass extinctions and change abiotic features permanently.

# Kingdom Monera

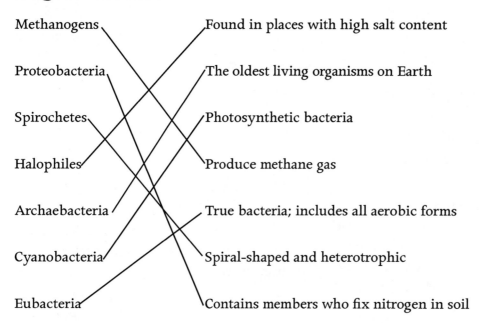

## Monerans and Disease

1. Monerans enter, overwhelm the immune system, and reproduce, causing systemic infection.
2. Sir Alexander Fleming cultured a dish of bacteria and, by accident, contaminated it with a mold spore. The mold produced penicillin, which killed the bacteria.
3. Bacteria may become resistant.

## Moneran Adaptation

1. a hard outer coating that protects a moneran in hostile environments or conditions
2. Moneran metabolism differences allow monerans to live in environments in which other organisms cannot survive.

## Kingdom Protoctista

1. The protoctists do not have a set of characteristics in common. Members of the kingdom are placed there if they do not conform to the characteristics of the other kingdoms.
2. protozoans, algae, and slime and water molds
3. Protoctists are composed of eukaryotic cells.

## Protozoans

1. Amoeba move by extending a cytoplasm "foot," moving toward it, then sending out the foot again. The foot, called a pseudopodia, encircles food sources and pulls them into the amoeba body.
2. malaria, yellow fever, giardiasis (Malaria and yellow fever can be controlled by keeping down mosquito populations; giardiasis can be eliminated by upgrading water-treatment plants.)
3. Flagellates move by using one or two whip-like tails. Ciliates move by coordinated movement of many tiny, hair-like appendages.

## Algae

1. Euglenoids have both animal and plant tendencies; diatoms are unicellular organisms with ornate and beautiful shells; and dinoflagellates can move on their own power through flagella.

2. Every time they reproduce asexually, they get smaller, since the shells do not grow. They must reproduce sexually to achieve larger offspring.

3. Kelp is the basis for an underwater ecosystem, which is home to many animals.

4. in the fur of sloths

## Slime Molds and Water Molds

1. They reproduce sexually, producing spores that are dispersed by the wind.

2. Cellular slime molds, unlike plasmodial slime molds, retain their own cell membranes.

3. Water molds kill root crops, such as potatoes.

## Ploidy, Meiosis, and Mitosis

1. Ploidy is the number of chromosomes in a cell. Sex cells are haploid.

2. There are four different phases of mitosis. The first, the prophase, is characterized by the DNA chromatin coiling up into visible chromosomes. Each duplicated chromosome is made up of two halves, called sister chromatids. They are exact copies of each other and are attached to each other by a membrane called the centromere. As the prophase continues, the nucleus seems to disappear. Two important structures—the centrioles, which are dark cylindrical structures—begin to move to opposite ends of the cell. Between them forms the spindle, a cage-like object. In the second phase, metaphase, the doubled chromosomes become attached to the spindle. They are pulled by the spindle fibers until they are neatly separated on the midline of the spindle. One sister chromatid of each chromosome is attached to one spindle fiber, while the other is attached to the other spindle fiber. In the third phase, anaphase, the separation of the sister chromatids begins. The centromere splits, allowing the two new chromosomes to separate fully. In the final phase, telophase, the chromatids have moved to opposite ends of the cell, following the centrioles. The spindle breaks down, and a new nucleus forms around each of the new chromosomes. Finally, the cytoplasm divides, and two new cells are formed.

3. In meiosis, four sex cells are produced. They must combine with a sex cell of the opposite type to create a new organism. Sex cells provide for genetic variability among a population.

## Kingdom Fungi

1. Most of the organism is underground.

2. They are important decomposers.

3. hyphae; mycelium

4. fragmentation, budding off, and sexual reproduction via spores

5. It is a spore-bearing structure that releases spores upon maturity. They can be produced by mitosis or meiosis.

## Zygospores and Sac Fungi

1. Drawings will vary, but should include landing on a good material, growth of

stolons and rhizoids, and development of sporangia.

2. Yeast are fungi, and many varieties are edible. Fungi can also cause disease in humans, animals, and plants.

## Club Fungi and the Deuteromycotes

1. The spores of the phylum Basidiomycota are produced in club-shaped hyphae called basidia. The mushroom typically consists of a large thallus, sometimes called a *stipe,* which supports a cap. Under the cap are *gills,* a series of membranes. Along the gills, the spores are contained in small club-shaped basidia. As the cap decays, wind, rain, or animal fur carries away the spores. When the spores land in a suitable environment, they germinate into hyphae, which grow down into the soil. Under the ground, a mycelium forms, with only one set of chromosomes. It is a haploid organism. The mycelium must mate with another organism to produce a diploid offspring. The two haploid cells fuse and form buttons—compact masses of hyphae, just below the surface of the soil. The buttons develop into mushrooms, and inside each basidium, the two haploid nuclei come together to form a diploid cell. Meiosis then occurs, and new nuclei are produced that become part of the spores.

2. They are the phylum to which penicillin belongs. They do not have a sexual reproduction phase.

## Fungi in the Nutrient Cycles

1. During decomposition, carbon-dioxide molecules are released to the air.

2. During decomposition, ammonia is produced.

3. Nutrients are limited and do not come from outer space.

## Kingdom Plantae

1. sample answers: multicellular, photosynthetic, cellulose cell wall

2. It keeps leaves from drying out.

3. Green algae plants have the same type of chlorophyll, and both produce cellulose cell walls.

4. no leaves, leaves that last only a day or two, thick coat of cuticle

5. vascular systems

## Photosynthesis

1. Six molecules each of water and carbon dioxide, in the presence of sunlight, become a glucose molecule and six oxygen molecules.

2. glucose

3. Chlorophyll usually has a green pigment.

4. Chlorophyll is a pigment; chloroplasts are the organelles in which it is stored in the plant.

## Spore-Bearing Plants

1. Nonvascular plants transfer water and sugars via osmosis; vascular plants use specialized cells. Nonvascular plants are usually very small and live close to water.

2. The sporophyte generation produces two structures for sexual reproduction, the male antheridium and the female archegonium. These two structures are the gametophyte generation, which must join together to create a new sporophyte organism.

3. They live near water. Sperm must move through water to reach the ova, so water is essential for sexual reproduction.

## Gymnosperms

1. naked seed
2. sporophyte generation
3. The fire clears out underbrush and leaves a thick layer of ash, which is a natural fertilizer. Fire warms the cones, and they open, dropping the seeds into this environment.
4. Cycadophyta, Gnetophyta, and Ginkgophyta
5. They lost ground to the angiosperms and to a drier environment.

## Angiosperms

1. flowers and fruit
2. possible answers: during the Mesozoic era, during the Jurassic period, 140 million years ago
3. annual—plant germinates, grows, flowers, sets fruit and dies in one year
   biennial—plant germinates and grows first year, flowers, sets fruit, and dies second year
   perennial—plant lives many years, flowers, and sets fruit on regular basis

## Kingdom Animalia

1. the blastula stage, during embryonic development
2. level of organization, symmetry, and body plan
3. Cellular—no organs or tissues; all processes take place within cells; tissue—some specialized tissues, but no organs or systems; organ—some specialized organs, but no communication between organs; system—groups of organs and tissues that function together to perform a necessary task
4. In bilateral symmetry, animals are mirror images around a central bisecting line. In radial symmetry, the animals are symmetrical around a central disk.
5. Humans have a head, a chest cavity with vital organs, an abdominal cavity with intestines and reproductive organs, and two arms and two legs.

## Marine and Aquatic Invertebrates

| E |   | O | C | T | O | P | U | S |
|---|---|---|---|---|---|---|---|---|
| C |   | B |   | N |   |   |   | E |
| H |   | I |   | N |   | P |   | S |
| I |   | L |   | D |   | O |   | S |
| N |   | A |   | A |   | R |   | I |
| O |   | T |   | R | A | D | I | A | L |
| D |   | E |   | I |   | F |   | E |
| E |   | R |   | A |   | E |   |   |
| R |   | A |   |   |   | R |   |   |
| M | O | L | L | U | S | C | A |   |

## Terrestrial Invertebrates

1. sample answers: pinworm infestation, trichinosis
2. all the way to the center of the animal
3. annelid

## Arthropods

1. jointed appendages, which are used for walking, carrying food, sensing the world, and mating

2. Exoskeletons are very protective. They are also very heavy, and must be shed from time to time because they do not grow.

3. Gills are used by aquatic species, because the species live mainly underwater and must get oxygen through osmosis. Book lungs are used by arachnids. Book lungs do not require muscular movement, which might give a spider or its web away. Insects have tracheal tubes. Insects move much more and fly, which requires greater oxygen levels in the blood.

4. Arthropod systems are simpler than higher vertebrate systems, at each organ in the system. However, they, like the higher vertebrate systems, communicate with other systems.

## Sea Squirts and Invertebrate Chordates

1. Phylum Chordata

2. The gills reappear on the inside of the siphon, and the animal creates a minor current in the water, which causes oxygen-rich water to pass over the gills.

## Fish

1. Jawless fish are either parasites or scavengers. Cartilaginous fish are true predators, such as the shark, with jaws that can open and close.

2. The lateral line is a line of several fluid-filled pits on the side of the animal's body. These aid the animal in sensing movement around them.

3. Each pair of fins does something different. Some keep the animal upright, some move the fish forward in the water, and some assist in steering and internal fertilization.

4. It allows fish to live at various depths, which has increased the number of niches available to them.

## Amphibians

1. intense competition for resources; land had plenty of food and oxygen, but also unequal air temperatures, which caused problems.

2. The young is born with gills, no legs, and a long tail. Over time, the legs grow, the tail shortens or disappears altogether, and lungs develop, which allows the amphibian to live out of water.

3. Amphibians obtain oxygen through their wet skin, and they must lay eggs in water.

## Reptiles

1. Answers will vary; sample answer: eyelids allowed the animal to exist in dry, sandy environments by allowing it to close its eyes during dust or sandstorms without damaging them.

2. Reptiles must bask in the sun for a period of time each day. This means that they must either live in warm areas year round, or must hibernate during winter months. It also means that they do not need as much sustenance, since ectothermic metabolisms do not need as much food as endothermic ones.

3. Garter snakes keep eggs inside their bodies until they hatch and care for a brood of young until they are nearly adults. Alligators guard nests, remove babies one by one, and carry them to the water.

## Birds

1. It is believed that birds descended from small theropod dinosaurs about 100 million years ago.

2. hollow bones, feathers, and endothermy

3. The female chooses the mate with the best possible characteristics, which will appear in her chicks. The mating ritual is the means by which the male shows the female his attributes.

## Mammals

1. Monotremes lay eggs; marsupials give birth to undeveloped young who must live within the pouch during development; placentals give birth to more or less fully developed organisms.

2. It is probably a small member of the same family to which horses belong. It has flat teeth for eating the abrasive grasses of the grasslands.

3. Answers will vary, but should include a protocol for reward for behavior.

## Defining Evolution

1. natural selection
2. sexual selection
3. sexual selection
4. natural selection

## Precambrian Time

1. Cyanobacteria; by producing oxygen, they wiped out many of the anaerobic life forms.

2. It is a formation of cyanobacteria fossils. By its position in the crust, the date for cyanobacterial life can be pinpointed.

3. Precambrian time lasted from the formation of Earth until 545 million years ago. The Earth was too hot for liquid water until 3.9 million years ago. Soft-bodied animals, some algae, and some protoctists existed at the very end.

## The Paleozoic Era

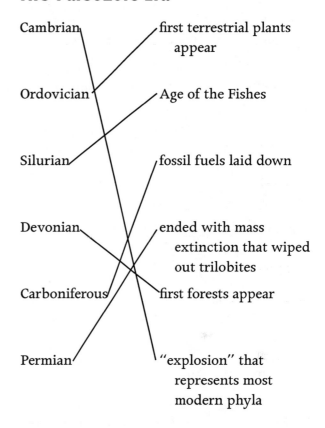

## The Mesozoic Era

1. Triassic, Jurassic, Cretaceous
2. the dinosaurs
3. mammals, flowering plants, and the first birds
4. flowering plants created many new niches for mammal species
5. An asteroid or large meteor struck Earth near the Yucatan Peninsula. There is a crater at the site, as well as a worldwide

iridium layer at that point in the fossil record.

## The Cenozoic Era

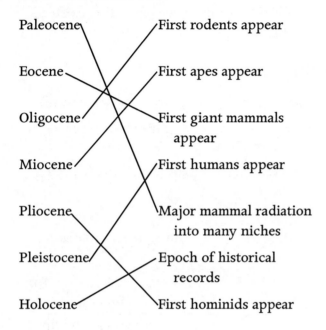

Paleocene — First rodents appear
Eocene — First apes appear
Oligocene — First giant mammals appear
Miocene — First humans appear
Pliocene — Major mammal radiation into many niches
Pleistocene — Epoch of historical records
Holocene — First hominids appear

## Human Evolution: *Homo*

1. *Homo habilis*

2. Gracile forms have thin bones; robust have heavy bones. Modern humans are gracile.

3. Neanderthals were robust forms who died out 28,000 years ago without leaving descendants. We are much more closely related to early *H. sapiens,* such as Cro-Magnon, than we are to the Neanderthals, even though they occupied Europe at the same time.

*Top Shelf Science: Biology* APPENDIX I

# Rubrics: Assessing Laboratory Reports

This book contains several student laboratory assignments for which you will produce written reports. Lab reports are important because they are a written recipe for another scientist to replicate your findings. Information should include:

- *Purpose:* Why is this lab being performed? What is the objective of the lab?
- *Hypothesis:* Given the initial level of knowledge, what do you expect to find out at the end?
- *Materials list:* A well-organized materials list makes it easier for anyone trying to replicate your results to understand what you did.
- *Procedure:* Even though a procedure is suggested in the lab write-ups, you should include the procedure you actually followed.
- *Data:* What actually took place in the lab?
- *Conclusion:* What were the results? Did your hypothesis match the data? If something went wrong, what do you think happened?

Below is a rubric your teacher may use to assess your lab reports.

|  | 1 | 2 | 3 | 4 |
|---|---|---|---|---|
| Understanding of concept | poor | adequate | good | outstanding |
| Methodology | poor | adequate | good | outstanding |
| Organization of experiment | poor | adequate | good | outstanding |
| Organization of report | poor | adequate | good | outstanding |

# Rubrics: Assessing Essays

In addition to lab reports, you will also be assigned several essays. You should take this opportunity to do outside research on the subject matter to supplement the material provided in this book.

Below is a rubric your teacher may use to assess your essays.

|  | 1 | 2 | 3 | 4 |
|---|---|---|---|---|
| Quality of research | poor | adequate | good | outstanding |
| Organization of material | poor | adequate | good | outstanding |
| Presentation of material | poor | adequate | good | outstanding |
| Spelling, grammar, and style | poor | adequate | good | outstanding |

# Scientific Supplies and Suppliers

Here is a short list of reliable scientific suppliers who will sell to home and school markets.

Edmund Scientifics
60 Pearce Ave.
Tonawanda, NY 14150-6711
800-728-6999
800-828-3299 (fax)
www.scientifics.com
Biology, earth science, chemicals (after providing forms)

Fisher Science Education
2000 Park Lane
Pittsburgh, PA 15275
800-955-1177
www.fisheredu.com  or  www.fishersci.com
Biology, earth science, physics, labware

Educational Innovations, Inc.
362 Main Ave.
Norwalk, CT 06851
888-912-7474
203-229-0740 (fax)
www.teachersource.com
Biology, earth science, physics, labware

Frey Scientific
100 Paragon Pkwy
Mansfield, OH 44903
800-225-3739
www.freyscientific.com
Biology, earth science, physics, chemicals, labware

For this book, the following materials are needed for lab exercises:

## Symbiotic Lichen:
- Samples of fruticose lichen
- Microscope, slide, and cover slip

## Bacterial Shapes and Arrangement:
- Microscope
- Prepared slides (or pond water, slide, cover slip, and iodine stain)

## Growing Water Molds:
- Small bit of uncooked hamburger or other meat
- Plastic container full of water
- Microscope, slide, and cover slip
- Iodine stain

## Observing Cell Division in Onions:
- Microscope, slide, and cover slip
- Onion root tips
- Iodine stain

## Bread Mold:
- Bread (preferably a kind with no preservatives)
- Substrate material (such as soil)
- Sandwich bags with a tie or zipper lock
- Marker to label the bags
- Damp paper towels

## Tree Rings:
- A tree section or tree stump containing the entire diameter of the tree
- Crayons and paper

## How Fish Swim:

- Fish tank with several species, trip to local aquarium, or nature video featuring fish
- Field guide to identify fish species

## Feathers:

- Microscope or hand lens
- Various bird feathers of all types

## Creating "Life":

- Microscope, slide, and cover slip
- Labware, including 50-mL graduated cylinder, 50-mL Erlenmeyer flasks (2), and a 400-mL beaker. Heat-resistant glass can be substituted.
- Gram scale
- 1% NaCl solution
- The following amino acids: aspartic acid, glutamic acid, and glycine. You can probably get these from a friendly chemistry teacher—none of them is dangerous. But you do NOT want to spill any of these on your clothes—the makings of life do not smell pretty!
- Hot plate or kitchen stove
- Ring stand and clamp
- Tongs
- Watch or clock with second hand
- Stirring rod
- Pipette
- Apron, oven mitts, and goggles

# A Time Line of Biology

1348—The plague appeared in Europe.

1508–1510—Leonardo da Vinci compiled notebooks on mechanics, astronomy, anatomy, and his inventions.

1628—William Harvey published a book describing blood circulation.

1660—Francesco Redi disproved theory of spontaneous generation with an experiment on flies.

1665—Robert Hooke invented the first microscope.

1673—Antonie van Leeuwenhoek saw bacteria for the first time.

1758—Carolus Linnaeus developed taxonomy of species and proposed binomial nomenclature.

1796—Edward Jenner discovered smallpox vaccination.

1838—Cell theory proposed by Matthias Jakob Schleiden

1856—Neanderthal fossil was found in Germany.

1859—Charles Darwin published *On the Origin of Species*.

1866—Gregor Mendel wrote a paper on his findings about heredity in plants.

1868—Cro-Magnon fossil was found in France.

1870—Process of mitosis observed for the first time

1890—Process of meiosis observed

1891—"Java Man" discovered in Indonesia

1900—Carlos Finlay discovered that yellow fever is spread by mosquitoes.

1902—Discovery of *Tyrannosaurus rex*

1903—The botanist Hugo de Vries discovered mutations in plants.

1904—Thomas Morgan performed genetic experiments that discovered sex-linked mutations (among a group of fruit flies with normal red or unusual white eyes, all of the white-eyed offspring were male).

1909—The "Piltdown Man" hoax—a fake archaeological discovery announced by dishonest scientists who wanted to prove that human beings had evolved in Europe

1923—Development of the diphtheria vaccine

1923—Production of insulin to treat diabetes

1924—Discovery of *Australopithecus africanus*; its human-sized brain too large to be that of an ape, but found to have the canine teeth of a gorilla

1925—John Scopes fired from biology teaching position for teaching evolution

1939—Discovery of Kirlian photography—electrical "auras" surrounding living specimens

1947—W. F. Libby invented radiocarbon dating.

1951—Rosalind Franklin discovered the helical shape of nucleic acids (RNA and DNA).

1953—James Watson and Francis Crick discovered DNA has double helix, composed of ATCG bases occurring in pairs (A with T, and C with G).

1953—Stanley Miller produced amino acids from inorganic compounds and sparks.

1953—Radioactive fluorine dating proved that the "Piltdown Man" artifact was a hoax.

1969—Meteorite in Australia found to contain amino acids

1972—Discovery of a 2-million-year-old humanlike fossil, *Homo habilis,* in Africa

1974—Discovery of "Lucy" in Africa, an almost complete hominid skeleton over 3 million years old; only 1 meter tall but found to have adult teeth, a small brain, and the ability to walk upright

1977—Submarine *Alvin* explored midoceanic ridge and discovered chemosynthetic life.

1979—First "test-tube baby" from artificial insemination

1991—Discovery of the buried crater near the Yucatan Peninsula, dated at 65 million years old

1996—Pope John Paul II affirmed evolution by natural selection.

1997—Microscopic analysis of meteorite led to belief in ancient life on Mars.

2000—Human genome mapped

# GLOSSARY

**angiosperms:** flowering plants

**amniotic egg:** egg with a membrane surrounding the fetus that contains amniotic fluid

**ATP:** adenosine triphosphate; energy source of the cell

**biome:** major ecological coummunity type, such as a desert or a tropical rainforest

**blastula:** hollow sphere of cells in the embryonic stage that later forms the alimentary canal

**cartilage:** bone-like connective tissue

**cell wall:** in plants, fungi, and some bacteria, a firm, inflexible structure that surrounds the plasma membrane; the cell wall provides structure and protection to organisms.

**centrioles:** cell organelles forming the spindle during cell division

**centromere:** the central area of a chromosome

**chlorophyll:** green pigment that traps light energy from the sun and gives plants and some protoctists their greenish color; other colors of chlorophyll, such as red, brown, and blue-green, also exist.

**chloroplasts:** chlorophyll-containing organelles found in green plants and algae

**chromatin:** thread-like structures of protein and DNA

**chromosomes:** structures of genetic expression

**cilia:** hair-like appendages for movement or sensory functions

**coelom:** body cavity, usually abdominal

**commensalism:** a form of symbiosis in which one partner benefits from the relationship and the other is unharmed

**cytoplasm:** clear fluid in eukaryotic cells surrounding the nucleus and organelles; the cytoplasm is the site of many important chemical reactions.

**cytoskeleton:** structural proteins of cytoplasm arranged as filiments

**dicots:** plants, such as strawberries, that have two seed leaves

**diffuse:** movement of particles from an area of high concentration to an area of low concentration, resulting in equal distribution

**diploid:** having two of each chromosome or twice the number of chromosomes that are present in sperm or ova

**ecosystem:** populations in a community and the abiotic factors with which they interact; for instance, a marine ecosystem

**endoplasmic reticulum:** organelle made up of canals, located in the cytoplasm of a cell

**exoskeleton:** an external supporting covering of an animal, usually one that does not contain vertebrae

**flagella:** whip-like structures that assist cells in movement

**gamete:** haploid sex cell

**Golgi apparatus:** sacs within the cytoplasm of cells that concentrate and package proteins for secretion from the cell

**haploid:** cell that has one of each chromosome type or half the number of chromosomes present in most other body cells

**hyphae:** thread-like filaments that form the basic structure of fungi

**limiting factor:** factor that necessarily limits the growth of populations, such as the amount of water, temperature, or number of prey species

**lysosomes:** organelles responsible for intracellular digestion

**meiosis:** cell division by which gametes or haploid sex cells are formed

**mitochondria:** cytoplasmic organelles that produce most of the cellular energy

**mitosis:** cell division that results in two identical daughter cells

**monocots:** plants, such as grasses, that have only one seed leaf

**mutualism:** type of symbiosis in which both partners benefit

**mycelium:** in fungi, the complex mass of branching hyphae; the mycelium is often the visible part of the fungus.

**notochord:** rod of tissue functioning as a spinal chord in embryos and some organisms

**nuclear envelope:** structure that surrounds the nucleus; it has pores that allow the transport of nucleotides (sugars used by the nucleus) and proteins into the nucleus, and ribosomes out of the nucleus.

**nucleolus:** site where ribosomal RNA is produced

**nucleus:** body within a cell containing DNA

**organelle:** structure of a cell with a specific function

**parasitism:** type of symbiosis in which one partner benefits at the expense of another

**petals:** leaf-like flower organs, usually brightly colored, that attract pollinators through color, perfume, or nectar production

**photosynthesis:** process by which species produce simple sugars from water and carbon dioxide in the presence of sunlight

**pistil:** in flowers, the female reproductive structure attached to the top of the flower stem; the lower portion of the pistil forms the ovary.

**plasma membrane:** boundary between the cell and the external environment; it allows the passage of nutrients and oxygen into the cell, and waste products out of the cell.

**ploidy:** number of chromosome sets in a living organism; most animals, for example, are diploid, having two sets of chromosomes. Sex cells, with only one set of chromosomes, are haploid.

**ribosome:** cytoplasmic organelle composed of protein and RNA where protein synthesis occurs

**sepals:** leaf-like structures at the base of flowers

**sister chromatids:** identical halves of the duplicated parent chromosome formed before the onset of cell division

**spindle:** fiber structure that forms between the two poles or centrioles during the prophase stage of cell division and shortens during anaphase, pulling the sister chromatids apart

**spindle fibers:** filaments that extend and attach to the chromosomes during the metaphase stage of cell division

**spores:** reproductive cells that can stem from either sexual or asexual reproduction; spores have hard outer coatings and are often microscopic. Some species reproduce by spores exclusively; others reproduce by a combination of spores and seeds.

**stamen:** in flowers, the male reproductive structure, consisting of an anther and attaching filament

**vacuole:** small space or hollow organelle within the cytoplasm of a cell

**zygote:** fertilized egg cell formed by the union of a sperm cell and egg

## A

abiotic factors, 5, 6
    effect of ice ages on, 16
    effects of long-term drought on, 15
abscission, 90
aerobes, 18
Age of Dinosaurs, 159
Age of Fishes, 156
Age of Invertebrates, 155
Age of Mammals, 163–164
AIDS, 3
albumen, 128
algae, 5, 32, 37–38, 71, 155, 156
allantois, 128
alleles, 47–48
alternating generations, 69, 74
amnion, 128
amniotes, 128–129
amniotic eggs, 117, 128–129, 131, 136
amoebas, 31
amphibians, 117, 124–126
    three major orders of, 125
anaerobes, 18–19
angiosperms, 77, 84–86, 164
    basic life cycles, 85
    fertilization, 84
    importance as food sources, 86
    increase of, 81
animals
    characteristics of, 95
    organization levels, 96
    symmetry and body plan, 97
annelids, 105, 111
annuals, 85
Antarctic Ocean, 28
antheridium, 74, 77
antibiotics, 25
    negative effects of overuse of, 26
    use of fungi for, 56
arachnids, 112

archaebacteria, 18–19, 152
*Archaeopteryx*, 160
archegonium, 74, 77
arthropods, 107–109, 155, 156
    evolution and diversity of, 111–113
    organization, symmetry, and systems of, 108
ascospores, 59
ascus, 59
asexual reproduction, 31, 57, 59, 99
ATP (adenosine triphosphate), 44, 51
autotrophs, 13, 37, 71

## B

bacteria, 2, 155
    decomposition and, 63–64
    drug-resistant strains of, 26
    Gram-positive, 19
    photosynthesizing, 5, 148
    shapes and arrangement (student lab), 21
basidia, 61
Bering Land Bridge, 16
biennials, 85
bilateral symmetry, 97
binary fission, 23
biology, 1–3
    definition of, 1
    time line of, 183–184
bioluminescence, 38
biome, 5
biotic factors, 5
    effect of ice ages on, 16
    effects of long-term drought on, 15
birds, 117, 136–137
    feathers (student lab), 139
    metabolism of, 137
    types of feathers of, 136
blastospores, 95
blastula, 95
bones, 117, 121
    middle-ear, 141
botulism, 29

bryophytes, 74
budding, 38, 46, 57
Burgess Shale, 155
burrs, 84
buttons, 61

## C

Cambrian period, 152, 153, 155
carbon, 63
Carboniferous period, 75, 156
carnivores, 12, 100, 102, 137
cartilage, 119–120
cell division
    meiosis and mitosis, 50–52
    observing, in onions (student lab), 54
cells
    characteristics of sexually reproducing, 50
    diploid, 50
    haploid, 50, 52
    specialization of, 99
    structure of eukaryotic, 43–44
cell walls, 44, 67
    fungi, 56
Cenozoic era, 145, 163–164
centrioles, 50, 51
centromere, 50, 52
cephalopods, 100–101
chitin, 56
chlorophyll, 44, 71
    kinds of, 37
    manufactured by cyanobacteria, 22
chloroplasts, 22, 44, 71–72
chordates
    invertebrate, 115
    vertebrate, 117
chorion, 129
chromatids, 50–51
chromatin, 43
chromosomes, 22, 46, 47, 50, 61, 77
cilia, 31, 115
ciliates, 35
cladistics, 2

class, 1
classification, 1–2
club fungi, 61
club mosses, 75
clutches, 134
coelenterates, 99–100
coelom, 95, 96
commensalism, 8–9, 111
competition, 5–6
conditioning, 144
cones, 77, 80–81
conifers, 80–81, 157, 159
constriction, 18
consumers, 12–13
    of an ecosystem, 63
corolla, 92
cortex, 88
cotyledons, 80
Cretaceous period, 160–161
cross-fertilization, 92
crossing over, 51
cross-pollination, 46–47
crustaceans, 112–113
cuticle, 67, 68, 105
cyanobacteria, 18, 22, 71, 152, 153, 155
cytoplasm, 18, 40, 43
    "foot" of protozoans, 34
    moneran, 22
cytoskeleton, 44

## D

Darwin, Charles, 115
Dead Sea, 19, 28
decomposers, 59, 63, 105
    role in ecosystem of, 13
decomposition, 55
    as last link in food chain, 13
deuteromycotes, 61
deuterostomes, 95–96
Devonian period, 117, 156
diatoms, 37, 156
dicots, 85

diffusion, 56, 125
dinoflagellates, 37–38
dinosaurs, 145, 157, 159–161
diploid
    cells, 50
    offspring, 61
diseases
    caused by bacteria, 19, 25
    caused by protozoans, 34, 35
    caused by sac fungi, 59
    caused by viruses, 3
    monerans and, 25–26
    serious plant, 40
    ticks as vectors for human, 112
DNA (deoxyribonucleic acid), 2, 43
    mitochondrial (mDNA), 44
    moneran, 18, 22
    protoctist, 31
    survey of sea squirts, 115
dorsal nerve cord, 115
downy mildew, 40
Dutch elm disease, 59

# E

Earth
    during Cenozoic era, 163–164
    history of, as calendar, 152–153
    important changes on, during Mesozoic era, 159
    major impact affects, 161
    origination of life on, 150
    oxygen in atmosphere of early, 18
echinoderms, 101, 155
*E. coli,* 23, 25
ecology
    definition of, 5
ecosystems, 5–6
    consumers of, 63
    importance of amphibians to, 126
    importance of fungi to, 55
    kelp forest, 38
    larger patterns in, 15–16
    vital role of scavengers and decomposers in, 13
ectothermy, 124, 131–132
embryos, 128, 129
endodermis, 88
endoplasmic reticulum, 43
endospores, 28–29
endothermy, 117, 136–137, 140
enzymes, 2, 56
epidermis, 88
estrus, 142
ethical issues, vii
eubacteria, 19, 21
euglenoids, 37
eukaryotes, 31, 119, 148, 153
    cell structure, 43–44
evolution
    bones, as advance in, 121
    defining, 147–148
    human: *Homo,* 168
    of true hominids, 166
exoskeletons, 56, 107, 117
exothermy, 117

# F

family, 1
fauna, 5
ferns, 75, 86, 159
fertilization
    amphibian, 125
    in birds and reptiles, 136
    fish, 121
    internal, 128
    plant, 84, 93
fins, 120, 121, 123
first-order consumers, 12, 13
fish, 120
    swimming and (student lab), 123
    three classes of, 119–121
flagella, 22, 35
    dinoflagellate, 37
flatworms, 101–102

Fleming, Sir Alexander, 26
flora, 5
flowers, 84, 92–93
    basic structure of, 92
    complete and incomplete, 92
food chains, 12–13, 63
food webs, 13
foraminiferans, 35
fossils, 152, 153, 156, 159, 160, 161, 168
fronds, 75
fruits, 77, 78, 84, 92–93
fungi, 5
    classification of, 57
    club, 61
    creation of penicillin from, 26
    importance of, in nutrient cycles, 63–64
    life cycle and structure of, 55–57
    sac, 59

## G

gametes, 47, 52, 69
    plant, 68
gametophytes, 69, 74, 75, 77
ganglia, 108
gene pool, 147
genes, 52, 147
    homozygous and heterozygous, 48
genetic code, 2, 46, 51
genetics, 46–48. *See also* meiosis; mitosis
genus, 1
germination, 61, 77, 93
Gila monster, 132
gills
    arthropod, 108
    fish, 119, 121
    mushroom, 61
    sea squirt, 115
*Ginkgo biloba*, 81
Golgi apparatus, 44
Gram's stain, 19
grana, 72
gravitropism, 86

Great Bombardment, 150, 152
Great Potato Famine, 40
Great Salt Lake, 19
guard cells, 89
gymnosperms, 77, 78, 80–81, 159
    decline of, 81

## H

halophiles, 19
Hanta virus, 3
haploid cells, 50, 52, 61
herbivores, 12, 159
heredity, 46–48
    development of modern laws of, 46
hermaphrodites, 102, 105
heterotrophs, 13
heterozygous, 48
holdfasts, 38
hominids
    evolution of true, 166
    use of tools by early, 166
*Homo sapiens*, 1, 168–169
homozygous, 48
hybrid, 48
hyphae, 11, 61
    club-shaped, 61
    of spores, 56, 57, 59
hypocotyl, 93

## I

ice ages, 16, 156, 164
immune system, 25, 28
indicator species, 6
infection, 25
insectivores, 12, 125, 132
insects, 95, 96, 108, 113
    successful body plan of, 97
invertebrates, 95, 115, 155, 159
    marine and aquatic, 99–102
    terrestrial, 104–105, 156
iridium, 161

## J

Jacobson's organ, 133
Java Man, 168
Jurassic period, 84, 159–160

## K

kelp, 38
Komodo dragon, 132

## L

larvae, 99, 100, 101, 113, 115, 125
leaves, 68, 84, 89–90
    adaptations of, 90
lichen, 147
    symbiotic (student lab), 11
life, kingdoms of, 1–2
limiting factor, 5
Linnaeus, Carolus, 1
lipid (fat) synthesis, 43
lysosomes, 44

## M

Malpighian tubule, 109
mammals, 117, 140–145, 163–164
    characteristics of, 140–141
    diet and adaptations, 141
    diversity of, 145
    learning and behavior, 143–144
    origins, 144–145
    reproduction and care of offspring, 142–143
mandibles, 109
mantle, 100, 101
marsupials, 143
mass extinctions, 16, 145, 148, 152, 156, 157
    caused by cyanobacteria, 18
    PT boundary, 159
medusas, 100
meiosis, 51–52, 61
Mendel, Gregor, 46–48
mesoderm, 95

Mesozoic era, 84, 111, 117, 145, 159–161
metamorphosis, 101, 113, 115, 125
methanogens, 19
Middle Ages, 59
mitochondria, 44
mitosis, 50–51, 52, 57, 93
mold, 59
    bread (student lab), 66
mollusks, 100–101, 155, 156
monerans, 18–19
    adaptation of, 28–29
    and disease, 25–26
    structure and reproduction, 22–23
monocots, 85
monotremes, 142
Montezuma's Revenge, 35
mosses, 69, 74
motility, 95, 115
Mount Saint Helens, 74
mutualism, 9, 11, 25, 92, 111
    of fungi, 55
mycelium, 55, 56, 57, 61

## N

National Science Education Standards, *vi*
natural selection, 147–148
nectar, 92–93, 137
nematodes, 104
Neolithic Age, 169
nitrogen, 63–64
nitrogen cycle, 63–64
notochord, 115
nuclear envelope, 43
nucleic acid, 2, 3
nucleolus, 43
nucleotides, 43
nucleus, 40, 43
    lack of, in monerans, 22
    number of chromosomes within, 50
nymphs, 113

## O

obligate aerobes, 28
obligate anaerobes, 28
omnivores, 12
order, 1
Ordovician period, 156
organelles, 43, 71
    lack of, in monerans, 22
organisms
    defense networks of, 25
    definition of living, 2
    environment of, 5
    photosynthetic, 18
    that provide energy, 12
organs, 96
osmosis, 18, 119
    in amphibians, 124–125
    in plants, 69, 74, 88
ova, 74, 77, 80, 99, 105
ovary, 84
    flower, 92
oxygen, 68
    as by-product of photosynthesis, 71
    monerans and, 28
    in primitive Earth's atmosphere, 18
    produced by algae, 32

## P

Paleolithic Age, 168
Paleozoic era, 75, 111, 117, 152, 153
    six major periods of, 155–157
Pangaea, 159
parasitism, 2, 3, 19, 111, 119
    of fungi, 55
    of protozoans, 34, 35
    types of, 8
parenchyma cells, 88
parthenogenesis, 109
pedipalps, 108
penicillin, 26, 61
perennials, 85
pericycle, 88
Permian period, 157
petals, 92
pH, 19
pheromones, 109
phloem, 69, 89
photosynthesis, 9, 63, 68, 71–72
    algae and, 37, 38
    equation for process of, 71
    simple sugars created through, 12
    in stems, 89
phototropism, 86
phylum, 1
pili, 22, 28
pistils, 92
placenta, 128, 143
plants, 156. *See also* seeds; spores
    dispersal of, 93
    dormancy of certain, 85
    herbaceous, 85
    history and adaptation of, 67–69
    spore-bearing, 74–75
plasma membrane, 44
plasmodium, 40
platyhelminths, 101–102
ploidy, 50
pollen, 80, 84, 93
pollination, 84, 92
polyploid cells, 50
positive reinforcement, 144
Precambrian time, 152–153, 155
precipitation, 63, 152
predators, 13, 159
primates, 163
producers, 12
prokaryotes, 43, 152. *See* monerans
proteobacteria, 19
prothallus, 75
protists. *See* protoctists
protoanimals, 99
protocell, 150

protoctists, 31–32, 155
  fungus-like, 40
protonema, 74
protostomes, 95
protozoans, 31, 34–35
pseudopodia, 31, 34, 35
psilophytes, 74
PT boundary, 157
pterophytes, 75
pupa, 113
purebred, 47

# R

rachis, 136
radial symmetry, 97
radiolarians, 35
recycling, 63
red tide, 38
regeneration, 101
reproduction
  algae, 38
  in angiosperms, 92
  of arthropods, 109
  of birds, 137
  bryophyte, 74
  and care of offspring of mammals, 142–143
  of ciliates, 35
  of diatoms, 37
  *E. coli*, 23
  of fungi, 57
  K and p reproductive strategies, 120–121, 125, 128, 140, 142
  mold, 40
  moneran, 18
  moneran structure and, 22–23
  protoctist, 31
  of reptiles, 134
  of sponges, 99
  of vertebrates, 117
  of viruses, 2
reptiles, 117, 131–134
  metabolism and feeding, 132–133
  reproduction of, 134
  sensory organs, 133
rhizoids, 59, 68
rhizomes, 75
ribosomes, 43
RNA (ribonucleic acid), 2
roots, 67–68, 84, 86, 88, 93
rubrics
  assessing essays, 180
  assessing laboratory reports, 179

# S

safety, *vii*
Sahara Desert, 15
scavengers, 102, 141
  role in ecosystem of, 13
sea squirts, 115
second-order consumers, 12
seeds, 80, 157
  disadvantages of, 78
  dispersal of, 84
  flower, 92, 93
  spores and, 77–78
segmentation, 97
self-pollination, 46
sepals, 92
septa, 56
sessile, 95, 99, 100
sexual selection, 147–148
sieve cells, 89
Silurian period, 156
siphons, 115
sister chromatids, 50
slime molds, 32, 40
spawning, 101, 121
species, 1
sperm, 46, 99, 105
sphenophytes, 75
spindle, 50, 51, 52
spindle fibers, 50
spiracles, 108

spirochetes, 19
sponges, 96, 99
sporangium, 57, 59
spores, 40, 56, 57, 69
   on bread, 59
   dispersal of, 61
   germination needs of, 77
   plants bearing, 74–75
   and seeds, 77–78
sporophytes, 38, 69, 74, 75
sporozoans, 34
stamens, 92
stems, 84, 86, 89
stereoscopic vision, 133
sternum, 137
stipe, 61
stolons, 59
stomata, 68, 89
Stonehenge, 169
*Streptococcus*, 21
stromatolites, 19, 22, 152
symbiosis, 8–9

## T

taproots, 88, 93
taxonomy, 1
teeth, 141, 142, 166
therapsids, 144–145
thermoacidophiles, 19
thermoregulation, 132
third-order consumers, 13
toxins
   effect on amphibians of, 126
   from endospore-forming bacteria, 29
   produced by dinoflagellates, 38
   released by bacteria, 25
   in wild mushrooms, 56
traits, 46, 147
   dominant and recessive, 47
transpiration, 63, 68

trees, 80–81, 147
   deciduous, 81, 90, 163
   tree rings (student lab), 83
Triassic period, 159
trilobites, 111, 155, 156, 157
trophic level, 13
tropisms, 86
tunicate, 115
*Tyrannosaurus rex*, 160

## V

vaccine, 25
   for tetanus, 29
vacuoles, 34, 44
   contractile, 35
vascular system, 68–69, 84
vectors, 34, 57
   ticks as, 112
vertebrates, 115, 156
   classes of, 117
vessel cells, 89
virion, 2
viruses, 2–3

## W

water molds, 32, 40
   growing (student lab), 42

## X

xylem, 68–69
   composition of, 89

## Y

yolk, 128

## Z

zooplankton, 100, 101, 115
zygospores, 59
zygotes, 47, 75, 77, 93

# Share Your Bright Ideas

## We want to hear from you!

Your name_____Date_____

School name_____

School address_____

City_____State_____Zip_____Phone number (_____)_____

Grade level(s) taught_____Subject area(s) taught_____

Where did you purchase this publication?_____

In what month do you purchase a majority of your supplements?_____

What moneys were used to purchase this product?

___School supplemental budget     ___Federal/state funding     ___Personal

**Please "grade" this Walch publication in the following areas:**

| | A | B | C | D |
|---|---|---|---|---|
| Quality of service you received when purchasing | A | B | C | D |
| Ease of use | A | B | C | D |
| Quality of content | A | B | C | D |
| Page layout | A | B | C | D |
| Organization of material | A | B | C | D |
| Suitability for grade level | A | B | C | D |
| Instructional value | A | B | C | D |

COMMENTS:_____
_____

What specific supplemental materials would help you meet your current—or future—instructional needs?
_____

Have you used other Walch publications? If so, which ones?_____

May we use your comments in upcoming communications?     ___Yes     ___No

Please **FAX** this completed form to **888-991-5755**, or mail it to

**Customer Service, J. Weston Walch, Publisher, P. O. Box 658, Portland, ME 04104-0658**

We will send you a **FREE GIFT** in appreciation of your feedback.  **THANK YOU!**